AFTER COAL

TOM HANSELL

AFTER COAL

STORIES OF SURVIVAL IN APPALACHIA AND WALES

West Virginia University Press
Morgantown 2018

Copyright © 2018 West Virginia University Press
All rights reserved
First edition published 2018 by West Virginia University Press
Printed in the United States of America

ISBN:
Paper 978-1-946684-55-4
Ebook 978-1-946684-56-1

Library of Congress Cataloging-in-Publication Data
Names: Hansell, Tom, author.
Title: After coal : stories of survival in appalachia and wales / Tom Hansell.
Other titles: After coal (Motion picture)
Description: First edition. | Morgantown : West Virginia University Press, 2018. |
    Includes bibliographical references and index.
Identifiers: LCCN 2018011995| ISBN 9781946684554 (pbk.) | ISBN 9781946684-561
    (ebook)
Subjects: LCSH: Coal mines and mining--Appalachian Region. | Coal mines and mining--
    Kentucky. | Coal mines and mining--Wales, South. | Appalachian Region--Economic
    conditions. | Kentucky--Economic conditions. | Wales, South--Economic conditions. |
    Appalachian Region--Social conditions. | Kentucky--Social conditions. | Wales, South--
    Social conditions.
Classification: LCC TN805.A5 H36 2018 | DDC 330.9429/4--dc23
LC record available at https://lccn.loc.gov/2018011995

Book and cover design by Than Saffel / WVU Press
Cover image: Women march in support of the miners during the 1984–85 strike. Photo by
Martin Shakeshaft (www.strike84.co.uk).

This book is dedicated to everyone who is not afraid to work for a better future.

In memory of
Terry Thomas (1938–2016)

CONTENTS

ACKNOWLEDGMENTS

First and foremost, I need to thank Pat Beaver, whose work on the exchange between Appalachia and Wales made *After Coal* possible. She believed in my capabilities and provided vital support through the Center for Appalachian Studies at Appalachian State University. I also owe a huge debt of gratitude to all of the After Coal project advisors, especially Helen Lewis and John Gaventa, whose work connecting Appalachia and Wales during the 1970s provided the foundation for *After Coal*, and Hywel Francis and Mair Francis, who introduced me to the Welsh valleys.

Billy Schumann, who became director of the Center for Appalachian Studies at Appalachian State University after Pat Beaver retired, also played a critical role in the production of this book. Additional support came from project advisors Ron Eller, Bill Ferris, Brian Winston, Richard Davies, Richard Greatrex, Frank Hennessy, Rich Kirby, Victoria Winckler, and Jack Wright.

Angela Wiley deserves particular mention. She started as an *After Coal* intern in the summer of 2012, and eventually worked her way up to becoming associate producer of the documentary and a contributor to this book.

A number of archivists and librarians supported my research, including Greta Browning, Fred Hay, and Trevor McKenzie from the W.L.

Eury Appalachian Collection at Appalachian State University; Sian Williams of the South Wales Miners' Library at Swansea University; Susan Williams and Elandria Williams at the Highlander Research and Education Center; Caroline Rubens and Elizabeth Barret at Appalshop's Archives; and Alexia Ault, who performed double duty at Appalshop and at the Appalachian Collection at Southeast Kentucky Community and Technical College.

I also need to thank my longtime friend Robert Gipe for welcoming me to Harlan County and providing access to the impressive network he has built there. Julie Bibby and Leslie Smith at the DOVE Workshop and Geraint Lewis and Big Al Lewis in Seven Sisters also provided important support. I am also deeply indebted to my wife, Sarah Carmichael, for her patience and support throughout this long-term project.

Content editor Leila Weinstein provided valuable feedback that helped strengthen this book considerably. Sandy Ballard, editor of the *Appalachian Journal*, offered invaluable guidance throughout this process. The Chorus Foundation and the West Virginia Humanities Council provided funding for the *After Coal* documentary, and this book would not have been possible without their support. Finally, I want to thank the staff at West Virginia University Press, especially director Derek Krissoff, for bringing this project to fruition.

Images from the production of the After Coal *documentary.*

INTRODUCTION

The After Coal project is part of a long-term exchange between the coalfields of central Appalachia and south Wales. Both regions lost thousands of mining jobs during the 1980s and 2000s, respectively. The After Coal project facilitated conversations between these two regions by producing a series of live public forums in coalfield communities, as well as a series of radio reports and a documentary film. The *After Coal* documentary has screened at international film festivals and on public television in the United States. This book provides an overview of the entire After Coal exchange through interviews conducted for the documentary, conversations from the live forums, and my personal reflections about making the *After Coal* documentary. Readers do not need to have seen the documentary in order to benefit from reading this book.

I tell this story in the first person, but an enormous team effort went into creating the After Coal project. I have tried to acknowledge everyone's contributions to this multifaceted project. Please forgive any oversights on my part. This exchange began long before I was involved, and I hope people from Appalachia, Wales, and other coal-mining regions will continue to come together to find common ground far into the future.

THE MOST DIFFICULT QUESTION

"EPA = Expanding Poverty in America."

This statement is written in three-foot-high letters on a banner stretched over a bandstand in a public park in Pikeville, Kentucky. It is June 2012 and I am just starting production of the *After Coal* documentary. The crowd around me is dressed in the reflective stripes of mining uniforms or in T-shirts reading "Friends of Coal" and "Walker Heavy Machinery." I am documenting a coal industry–sponsored pep

Images from the production of the After Coal *documentary.*

rally before a public hearing on new water-quality regulations proposed for mountaintop-removal coal mines.

The speaker onstage is speaking proudly of his family's heritage in the coal industry. He concludes his passionate statement with a question: "If we can't mine coal, what are we going to do in eastern Kentucky?"

Good question. As a filmmaker who has spent my career living and working in the coalfields of eastern Kentucky and documenting coal-mining issues, this is an important and difficult question to answer. My earlier documentaries *Coal Bucket Outlaw* (2002) and *The Electricity Fairy* (2010) were intended to start a civil conversation between workers in the coal industry and other community members about a shared vision for good jobs, clean air, clean water, and a safe working environment. However, the conversations almost always broke down as soon as someone pointed out the obvious: the coal industry had long been the only model of economic development in the central Appalachian region. More examples of what life after coal might look like were desperately needed to move the conversation forward.

As I struggled with the haunting question "If we can't mine coal, what are we going to do?" the image of Welsh mining villages rising from the ashes left by the coal industry captured my imagination. I thought that if I could just learn a few details about how Welsh communities made the transition, then I could identify specific solutions to help coal communities in Appalachia. However, I quickly learned that the secret to life after coal was not that simple.

FULL DISCLOSURE

The story of how I arrived in Wales is as long and winding as the mountain roads I traveled on both continents during the course of this project. However, I feel a responsibility to tell my personal story so that readers can better understand how I chose to approach the After Coal project.

When I was around nine years old, I became obsessed with fossils. These ancient life-forms preserved in rocks were magical, ancient, part of something larger than myself. I thought that if I looked hard enough, I would find the key that connected my young life to the ancient rhythms of the universe.

On my ninth birthday, my parents took me camping in the mountains of West Virginia. West Virginia was a great place to look for fossils. I had not walked far from my family's campsite before I found an abandoned coal mine. A massive slag heap towered over a group of block buildings wrapped in kudzu; while the discarded slate in the slag heap provided a rich source of fossils, I became fascinated by how beautiful and terrifying the abandoned mine was.

My boyhood obsession with fossils grew into an adult fascination with fossil fuel. Like the slag heap, the subject was beautiful and terrifying at the same time. I saw beauty in the mining equipment abandoned by the side of the roads I traveled on these family camping trips. The rusting hulks reminded me of dinosaurs: large, powerful, and ultimately doomed. These machines bore witness to the power of human ingenuity, but at the same time they acted as a terrifying reminder that our dependence on fossil fuels threatens our very survival.

My obsession with fossil fuels drew me away from my suburban home in Ohio to rural eastern Kentucky. I believe the Appalachian coalfields are a vital place to begin to address the problems associated with our dependence on fossil fuels. We burn an increasing quantity of coal, oil, and gas each year to meet the energy needs of an expanding population base. Science tells us we have a finite amount of fossil fuels on this planet. Together, these two facts present one of the major problems we must solve in order to sustain human existence for the long term.

In the Appalachian Mountains, this problem is strikingly evident. Coalfields developed a century ago are nearing the end of their productivity. Coal employment peaked around 1950 and has dropped steadily ever since (Energy and Environment Cabinet 2016). Remaining coal seams are increasingly difficult to mine, requiring radical techniques such as mountaintop removal. The local economy that depends on the extraction of fossil fuels has resulted in structural poverty—the percentage of the region's residents living under the poverty line has not improved over the past half century (Appalachian Regional Commission 2015, 6). If our society wants to wean itself from our dependence on fossil fuels, Appalachian communities deserve to be one of the first places where we seek solutions.

On my own quest for solutions, in 1990, I began my career at Appalshop, a rural, multidisciplinary arts center located in Whitesburg, Kentucky—the heart of the central Appalachian coalfields. From my young and naively privileged perspective, moving to eastern Kentucky was an act of opposition to the materialistic, consumer-driven world. I had a goal of living self-sufficiently, fulfilling my needs with what I could make or grow, and buying as little as possible. And, as an aspiring environmental activist, the clear moral lines around the issues in the Kentucky coalfields, especially strip mining, were appealing. The battle call of union songs such as "Which Side Are You On" charged up my little post-punk heart.

However, my experience at Appalshop quickly taught me that the struggles of coal communities were not as simple or straightforward as I had imagined. Working as part of this artistic collective, I produced radio and video documentaries and taught community media workshops. As a young artist and activist, I quickly absorbed Appalshop's mantra of providing a platform for mountain people to speak in their own words about issues that affect their lives. I attended hundreds of community meetings: school board, the fiscal court, mine permit hearings, and union meetings. I also documented dozens of direct actions where citizens blocked roads to stop mining, took over government offices to protest the lack of enforcement, and set up picket lines to enforce union contracts.

My experiences working on the front lines of the environmental justice movement in Appalachia gradually developed my understanding of the complexities of how culture, place, and politics had shaped the situations I was documenting. I witnessed firsthand the incredible power of community to support people as they faced threats against their homes and families. As a result, I expanded my ideas about self-sufficiency from an individualistic vision of each person taking care of their own needs to a larger vision of individuals living in symbiosis with their neighbors and the natural environment—community self-sufficiency.

Participating in cultural exchanges at Appalshop also provided me with valuable lessons. Meeting artists from the mountains of western China and rural Indonesia opened my eyes to some of the universal challenges faced by regional cultures in an increasingly globalized economy. I hoped that an international exchange with another coal-mining region such as south Wales could identify resources and strategies that would help Appalachian coalfield communities create a future beyond coal.

The process of creating the *After Coal* documentary took more than five years. During that time, I learned to stop looking for concrete solutions and start supporting an ongoing conversation about how to create healthy communities in former coal-mining regions. International efforts to address climate change make this challenge especially intense for coal-producing regions. As our economy shifts from fossil fuels, how can we ensure that places where fossil fuels were extracted do not continue to bear an unfair share of the costs of extraction?

I believe there are as many solutions for life after coal as there are residents of mining communities. I hope these stories from south Wales and central Appalachia will inspire people to discover solutions that work in their home communities.

An abandoned coal conveyor belt in Harlan County, Kentucky.

Painted rock above the valleys of south Wales reads "Cymru Am Byth" which translates to "Wales Forever."

A closed mine-supply warehouse in Harlan, Kentucky (above), is contrasted with a repurposed miners' hall in Ystradgynlais, Wales (below).

CHAPTER 1

WHY APPALACHIA AND WALES?

The industrialized coal-mining areas of South Wales and Central Appalachia share several common features: similar highland environments, a history of rural subsistence agro-pastoral economies, colonial experiences as a result of capitalist expansion, and industrialization based on extraction of minerals. Both regions have maintained viable subcultures and developed regional consciousness despite (or because of) their industrialization and integration into an international economy.

—Helen Lewis (Lewis 1984)

The mountains of Harlan County, Kentucky (left), and Afan Valley, Wales (right).

IN THE BEGINNING

"We need to turn off the lights in here."

I was shocked to hear this statement from the podium, as author Jeff Biggers looked out at a crowd of about three hundred students packed into the student union at Appalachian State University. Usually visiting speakers want to see their audiences. But I followed his instructions, walking over to the control panel to dim the lights until only the faint glow of the Exit signs lit the room.

"Whoops, I forgot I wouldn't be able to see my notes," Biggers quipped, and the audience laughed.

He paused then got serious: "Today is a great day for Chile. Thirty-three miners were trapped in complete darkness, darker than this, for more than two months. And thirty-three miners were able to emerge! Today, we celebrate with the people of Chile.

"And yet, our hearts were also broken today. In West Virginia, we lost another coal miner, who will not see his wife and family tonight. This is the forty-fifth American coal miner this year who will stay in darkness forever. . . . These lights are off for the families who lost men in the Upper Big Branch explosion this past April. We keep them in our hearts. When we turn on the lights, think of the cost of the coal we use to bring us out of the darkness."

In this way, news of the dramatic rescue of the Chilean miners on October 14, 2010, set the stage for the opening night of a symposium titled "Appalachia and Wales: Coal and After Coal." The three-day event was held at Appalachian State University in Boone, North Carolina. Pat Beaver, the director of the Center for Appalachian Studies on campus, and I, a faculty member working for the center, had organized a full agenda of presentations, workshops, and facilitated discussions about the common history of coal in central Appalachia and south Wales. Our idea was to invite community organizers from the Appalachian coalfields to meet with scholars and activists from Wales. The goal of the symposium was to share lessons and identify resources to help build a better future in coalfield communities worldwide. Although I was unaware of it at the time, this symposium was the genesis of the *After Coal* documentary.

APPALACHIA & WALES:
COAL AND AFTER COAL

Dr. Hywel Francis, MP
Wales

Dr. Helen Lewis
Appalachia, 1967

OCTOBER 14-16, 2010

**Table Rock Room of the
Plemmons Student Union
Appalachian State University**

FEATURING:

Jeff Biggers	Coal River Mountain Watch
Hywel Francis, MP	Kentuckians for the
Helen Lewis	Commonwealth
Mair Francis	Alliance for Appalachia
Ronald L. Lewis	Appalachian Voices
William Schumann	Ohio Valley Environmental Coalition
Amanda Starbuck	Southern Appalachian Mountain
Matthew Wasson	Stewards
Randy Wilson	ASU Appropriate Technology Program
Beth Bingman	Appalachian Institute for
Brad Nash	Renewable Energy
Patricia Beaver	Radical Roots Exhibit
Thomas S. Hansell	by Taylor Kirkland

Speakers' books available for purchase and signing through the Appalachian State University bookstore

Appalachia and Wales: Coal and After Coal is presented by the Center for Appalachian Studies at Appalachian State University with major financial support from the University Forum Committee, and the Office of Academic Affairs. This project is made possible by a grant from the North Carolina Humanities Council, a statewide nonprofit and affiliate of the National Endowment for the Humanities. Co-sponsors include: departments of Anthropology, Communication, Government and Justice Studies, History, Philosophy and Religion, Sociology, Sustainable Development, International Programs, the Institute for Energy, Economics, and Environment, University Documentary Film Services, and the Appalachian Journal.

The Center for Appalachian Studies is a unit within Appalachian's University College. University College consists of the university's integrated general education curriculum, academic support services, residential learning communities, interdisciplinary degree programs and co-curricular programming—all designed to support the work of students both inside and outside of the classroom.

NORTH CAROLINA
HUMANITIES
COUNCIL
MANY STORIES. ONE PEOPLE.

Appalachian
STATE UNIVERSITY

UNIVERSITY COLLEGE
Appalachian State University

Poster advertising the 2010 Wales and Appalachia Coal and After Coal symposium.

Many of the participants had firsthand experience with the negative impacts of the coal industry. Cracked foundations, orange streams full of acid mine drainage, cancer clusters, mud, and dust from coal hauling were part of their daily experiences. Yet, in the heart of coal country, calling out these injustices was seen as a threat to your neighbor's job. The coal industry dominated most aspects of life. The coalfield residents attending the symposium were looking for an alternative, for an answer to the recurring question: If we don't mine coal, what do we do?

Jeff Biggers had recently completed a book about his family's tempestuous relationship with coal titled *Reckoning at Eagle Creek*. His keynote speech detailed the ways in which the current economic system at play in the coalfields was broken. He explained that the choice between clean water and well-paying jobs was a false choice that divided communities and supported the efforts of outside corporations to transform the wealth of the coalfields into bounty for shareholders outside of the region.

Despite these stark realities, Biggers conveyed a sense of hope. He shared stories from his 2005 book, *The United States of Appalachia*, which describes the key roles Appalachian people have played in major social movements in US history, from the movement for independence from Great Britain to the abolitionist, labor, and civil rights movements. Biggers cited these examples to suggest that Appalachia can become a leader in a national effort to address climate change and create a fossil-free economy.

The speakers that followed over the three-day event built on these themes. Some described how the coal industry became a dominant force in central Appalachia and south Wales. Others discussed how residents in both regions have organized to improve their communities. On the third day, participants discussed successful organizing strategies and considered how they might be applied more broadly.

As symposium participants collectively tried to figure out ways to talk about a sustainable coalfield economy, we faced two major challenges. The first was steering conversations away from the well-worn path of the coal industry's abuse of land and people. Instead, we hoped to focus on potential solutions. The other challenge was how to reconcile the intense local need for a sustainable economy with global efforts to address climate change.

To meet these challenges, many of the speakers expressed the need for coal country to move beyond the standard definition of economic development. Martin Richards, from the grassroots group Kentuckians for the Commonwealth, summed up most current economic development efforts as "more of the same," as local and state government officials focused on attracting more extractive industries such as coal, timber, and natural gas to the region. He challenged participants to move beyond the idea of economic diversification, which is to "add a few things to what is already there," and instead embrace the concept of economic transition. Richards described economic transition as a complete restructuring of economy and community to create a sustainable future. As we explored the concept of economic transition, we hoped that Wales may have some answers. We looked across the Atlantic for proof that there can be life after coal.

The Welsh presenters included Hywel Francis and Mair Francis, a married couple who had participated in exchanges with Appalachia since 1974. A labor historian, Hywel Francis was serving as a member of parliament, and Mair had helped found the DOVE Workshop, which provided training opportunities and employment for women in former mining communities. Both Hywel and Mair were crucial to the success of the After Coal project; I will explain their roles in more depth in the coming chapters. To add another international perspective to the conversation, we invited Amanda Starbuck, who had Welsh roots but was currently living in the United States working for the Rainforest Action Network on coal-related issues.

After listening to the Welsh presentations, participants developed a set of similarities and differences between Appalachia and Wales— many of which we will explore more deeply in subsequent chapters of this book.

The main similarities between the coalfields of central Appalachia and south Wales identified at the symposium included:

- Both regions shifted quickly from agriculture to the coal industry in the nineteenth and twentieth centuries.

- Entire systems of both societies (economic, political, and cultural) centered around the coal industry.

- Both regions lost that industry.

- Both regions have strong local identities that set them apart from their respective nations.

- Both regions have developed social capital as residents have organized around labor, culture, and environment.

Symposium participants also defined important differences between the two regions.

- Geography: The coalfields of south Wales run less than fifty miles from east to west and approximately twenty miles from north to south, while the coalfields of Appalachia stretch more than five hundred miles between Alabama and Pennsylvania. The central Appalachian coalfields featured in *After Coal* are a much smaller area of eastern Kentucky, southwestern Virginia, and southern West Virginia, but this area is still about three times the size of south Wales.

- Government: The Appalachian coalfields span state lines, which means that efforts to support economic transition in the region require the coordination of several state governments. In Wales a single governing body, the National Assembly, works with the United Kingdom's central government in London to coordinate support for former mining communities.

- Culture: Although coal has been a central part of regional culture in both Wales and Appalachia, Welsh communities appear more unified than Appalachian communities. Perhaps this is due to the strength of the mine workers union in south Wales—the region was considered a closed shop, meaning that miners were required to join the union in order to work. Symposium participants from Appalachia described tensions between coal miners and environmentalists. Many environmental activists, or "tree huggers" in local parlance, described being threatened when they spoke out against destructive mining practices. People shared stories of being shouted down at public hearings and having their dogs shot as a result of their opposition to mountaintop-removal coal mining. Highlighting this difference, Mair Francis explained that, although communities in south Wales are not entirely free of conflict, "we don't have guns, so we don't shoot each other's dogs."

- Politics: The biggest differences between the regions can be traced to dissimilarities in national politics. In the United States both major parties are decidedly capitalist in orientation, while in the United Kingdom, the Labour Party has historically supported socialist policies—most importantly the nationalization of key industries, including coal. In the UK, the coal industry was nationalized from 1947–92, which means that many former mine lands are now public land managed by the Forestry Commission. Meanwhile, in the Appalachian coalfields, outside corporations control more than half the land.

- The trajectory of the decline of the coal industry in each nation highlights these differences in national politics. In the United

Kingdom, the coal industry declined after Margaret Thatcher's conservative government decided to privatize the industry, and private utilities increasingly chose low-cost imported coal to generate electricity. In the freewheeling energy markets of the United States, coal lost out to cheap natural gas produced by fracking. Stricter regulations on air and water pollution enforced by the Obama administration helped natural gas dominate US energy markets as many utilities retrofitted old coal-fired power plants to burn the cleaner fuel.

The question on everyone's mind at the forum was "where do we go from here?" More specifically, participants wanted to know: How do we create public space for dialogue? How do we build bridges across divisions in our communities? And how do we access resources to create a better future for ourselves?

These are the questions that stayed in my heart and mind as I started work on the *After Coal* documentary.

On the last day of the symposium, Amanda Starbuck of the Rainforest Action Network wrapped up her presentation by calling for a clean energy revolution, a vision that would require us to both resist and create. She challenged symposium participants to use existing community resources to build skills and a knowledge base that will help communities transition to a post-coal future: "We need to honor the past in these places. It took an enormous amount of ingenuity, creativity, and people power to make the industrial revolution. We need to learn from that, and be inspired by that, and do it again."

Although I agreed with Starbuck's commitment to clean energy, I felt that an important piece of the solution was missing. My experience living in the Appalachian coalfields taught me that clean energy is not enough. If we care about our communities and our neighbors, we need to look beyond the simple answer of investing in clean energy to develop solutions that take care of displaced workers in coal and related industries. Perhaps what we need is a post-industrial revolution, a new economic structure that allows us to escape the control of the robber barons that dominated the industrial age and the financial traders who dominate the information age. An economy that allows former coal communities to develop locally owned and controlled enterprises.

The symposium got me fired up. I wanted to more deeply explore what a post-coal economy might look like, and it seemed clear to me that the former coal communities in south Wales might yield some answers. I realized that a documentary might be the right way to delve into these issues. I consulted with Pat Beaver, who was the director of the Center for Appalachian Studies, and we started planning a trip to south Wales.

HISTORICAL CONTEXT

As I prepared to travel to the coalfields of south Wales, I realized how little I knew about the history of coal mining in other parts of the world. My work as a radio reporter had taught me that coal is found in more than thirty nations. Traveling to other coal-mining regions had taught me that Russia is the nation with the largest reserves, while China is the biggest producer and consumer of coal (IEA 2013). Of course, data on coal production and consumption changes as new

Miners load coal in Lynch, Kentucky.

nations industrialize and others lose industry. For example, about the time the *After Coal* documentary was released in 2016, India surpassed the United States as the world's second largest consumer of coal.

Although I had a basic grasp of current coal markets, I knew very little about the historic coalfields of the UK. It turns out that Great Britain was the world's largest producer of coal from the beginning of the industrial revolution until 1909 (News 2012).

The British Isles has more than a dozen separate coalfields, but the intense concentration of mines in the small area of south Wales sets the region apart. The rapid industrialization of this region led to coal dominating all aspects of life for almost two centuries. Another feature that set south Wales apart from the other coalfields in the British Isles was the militant nature of the labor movement in the area. Many union leaders were syndicalists, Marxists, and indeed communists, who believed that the coal industry should be owned and managed by the workers. This legacy continued among many union members—prompting Coal Board officials to jokingly refer to the South Wales Area headquarters of the National Union of Mineworkers (NUM) as "the little Kremlin" (Weeks 2004).

I started my research by reading *Welsh Americans: A History of Assimilation in the Coalfields* (2008), historian Ron Lewis's account of Welsh miners who immigrated to the United States between 1840 and 1870 to work in the coal industry. This fascinating history reminded me of a story my grandfather once told me about his grandfather (my great-great grandfather) who conversely migrated from Wales as a teenager in order to *escape* work in the coal mines. He eventually settled in Utica, New York, and married into a family that owned the local auto dealership, thus cementing his move to the middle class.

These stories led me deeper into the history of Wales. Along the way I read popular Welsh literature such as Richard Llewellyn's classic novel *How Green Was My Valley* (1939), which traces the social upheaval that accompanied the industrialization of the Welsh valleys in the nineteenth century through the stories of one family. I also devoured assorted local myths including legends about the wizard Merlin's origin in the Welsh moors and King Arthur's travels there. Many of these books presented Wales as a beautiful rural land lost in time, with little mention of the industrialization that dominated the coalfields in the nineteenth and twentieth centuries. The similarities between these accounts and historic depictions of the Appalachian region were striking.

As I continued my research about Wales, one major difference I noted was that in Appalachia I could speak with people who had directly witnessed most of the region's history with coal, which in parts of Appalachia such as Harlan County, Kentucky, began less than one hundred years ago. And almost everyone I met had stories from previous generations who remembered what the region was like prior to industrialization. Scholar and oral historian Alessandro Portelli observes, "In Harlan County you feel the nearness of the beginnings. The stories go back to a pristine wilderness, the first migrations and settlements, the Revolution, yet this is a living memory, entrusted to generations of storytellers" (Portelli 2012, 14). In contrast, the Welsh experience with coal goes back almost ten generations. As a result, I was forced to rely on published texts to provide information about the development of the Welsh coalfields. As I read these accounts, I got the sense that the industrial history of south Wales had created a coal-mining culture not unlike Appalachia's. This mining culture gave rise to a community spirit. Dai Francis, leader of the South Wales Area National Union of Mineworkers, summed it up:

> If you go around south Wales you'll see that there are very narrow valleys. The miners live together in small mining

villages. They marry girls from the same village. The children are brought up in the same village, in the same valley. The disasters they share together. The fatal accidents they share together. Because everyone knows one another in the mining valleys. It's a community in the right spirit (Lewis 1976a).

FAIR WARNING

I want to preface this historical overview by acknowledging that I am a filmmaker, not a historian. However, through my work documenting the Appalachian region, I have learned that an understanding of its history is essential to understanding current issues. Therefore, I chose to include a section providing historical context in both the film and this book so that viewers and readers can deepen their understanding of the historic and cultural resources available for regeneration in these former mining communities. The abbreviated history of the coalfields that follows is punctuated by interviews recorded during the production of the *After Coal* documentary to provide readers with firsthand accounts and perspectives.

If you already have a good grasp of the history of the coalfields in these regions, feel free to skip ahead to chapter 4, where strategies for regeneration are discussed. Or, if you are interested in exploring this history in greater depth, consider reading some of the works I relied on when writing this chapter. They include Ron Eller's *Miners, Millhands and Mountaineers* (1979) and *Uneven Ground* (2008), Ron Lewis's *Welsh Americans* (2008), and Alesandro Portelli's *They Say in Harlan County* (2010). For background about Wales I relied on Hywel Francis and Dai Smith's history of the South Wales Miners' Federation, simply titled *The Fed* (1998), as well as Hywel Francis's eyewitness account of the UK coal miners' strike in 1984, *History on Our Side* (2009). Ben Curtis's *The South Wales Miners, 1964–1985*

(2013) is also a useful resource for those interested in learning more about Welsh labor history.

WALES

The smokeless quality of Welsh bituminous steam coal was unmatched anywhere in the world until the southern West Virginia coalfields were developed at the turn of the twentieth century.

—Ron Lewis in *Welsh Americans* (2008)

INDUSTRIALIZATION

During the nineteenth century, the increased demand for coal to power railroads, steamships, and other steam engines pushed coal mining into the valleys of south Wales. Rural forests and farms were quickly transformed into densely populated mining villages with terraced houses climbing up the hillsides, and by the mid-nineteenth century the coal industry had become central to the Welsh economy.

In addition to coal, Wales held substantial reserves of copper, tin, and iron ore. These resources placed south Wales at the vanguard of the industrial revolution. The region supplied the raw materials that fueled Great Britain's rise to power on the world stage. The industrialization of south Wales drew a diverse labor force from rural parts of Wales and Great Britain, as well as across Ireland and western Europe, including a significant number of workers from Spain. By 1900 the region's population had exploded and more than one quarter of the south Wales workforce were coal miners. During the industry's peak, in 1914, more than 230,000 men were employed in 485 pits across south Wales (Davies 2007).

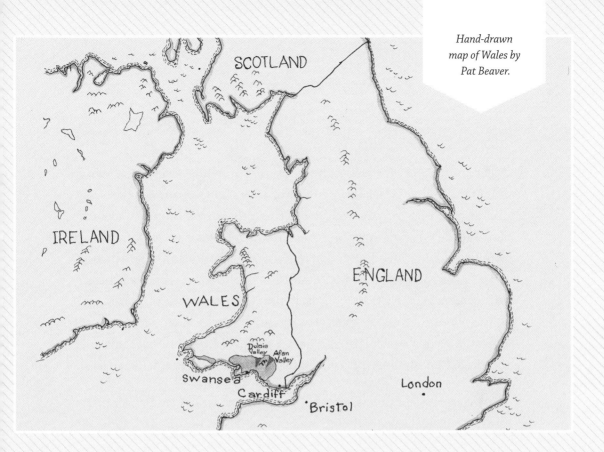

Hand-drawn map of Wales by Pat Beaver.

A view overlooking the Rhymney Valley in Wales.

ORGANIZED LABOR

Although coal mining is dangerous and difficult today, the early miners in south Wales labored under especially oppressive working conditions. Miners' pay was tied to the tons of coal they mined each day. The narrow twisting coal seams in south Wales meant that miners often worked twelve-hour days in dangerous conditions just to keep a roof over their head and food on their table. Safety and healthcare for workers was an afterthought, if any such provisions even existed. These conditions led workers to organize into local trade unions to demand a safer workplace and better wages. Miners' unions became an important element of community life in the valleys.

Historian Hywel Francis writes: "The earliest form of trade unionism in the South Wales coalfield, the Scotch Cattle, dominated some of

MERTHYR RISING

One of the first major labor battles in south Wales was the Merthyr Rising of 1831, when miners and iron workers took over the industrial town of Merthyr. Workers broke into the debtors prison, freed the inmates, and destroyed records of debts owed. The leaders of the uprising chose a red flag to symbolize the workers' blood that had been shed in the coal mines and iron foundries. This red flag later became an international symbol of workers' solidarity. The revolt that started at Merthyr quickly spread throughout the valleys of south Wales. At the urging

of the industrialists, the British Crown ordered the military to restore order in the valleys. Many soldiers and workers died in the ensuing conflict. After a week of rebellion, the military restored control to the ironmasters and mine owners. The leaders of the uprising were arrested and two men were sentenced to death. Only one man, Dic Penderyn (aka Richard Lewis), was eventually hanged, many think unjustly. He became a martyr, and the legacy of the Merthyr Rising continued to inspire the workers' movement in the decades that followed.

During a presentation at Southeast Community and Technical College in Harlan, Kentucky, Merthyr native Richard Davies provided background information about the development of industry in his hometown:

"**INDUSTRIALIZATION** came to Merthyr very early. Merthyr Tydfil industrialized about 1755, it industrialized with iron and steel. That steelmaking process used trees from our forests to make charcoal to smelt the iron ore. You began to get the development of small little communities the early decades of the nineteenth century . . . Anyone who broke ranks in a strike was ostracized in the community. The rough treatment of scabs could be interpreted as 'intimidation,' but so could the famine and distress imposed on communities by coalowners, iron-masters, and their judicial and military allies" (Francis 2009).

The Mines and Collieries Act of 1842 was the first British law to regulate coal mines (Parliament). Responding to the dangerous labor conditions, the law banned women and children from the mines but did not change conditions for the men who continued to labor under-ground. Miners continued to organize into local trade unions and to demand higher wages and safer working conditions. These small, local unions were well established by 1890. Following the bitter defeat of a miners' strike, ten small unions, called miners' associa-tions, banded together to create the South Wales Miners' Federation

MERTHYR RISING

around individual ironworks. People began to flood in by the thousands [for work] and, by 1800, Merthyr Tydfil became the biggest town in Wales. The next thing that happened is the ironworks begin to close in about 1880–90. This was the first stage of deindustrializa-tion of Merthyr Tydfil, but what replaced it was a second phase of industrialization—the coal industry.

"Welsh coal goes around the world and one of the centers was Merthyr Tydfil, which produced fine steam coal. The British Navy was fueled by Welsh coal. The British Empire was fueled by Welsh coal. The names of the pits, Great Western, the Naval Colliery, all indicated where the coal was used. Because south Wales was an export coalfield, its markets were broad. After the First World War, coal exports began to decline and a number of pits began to close.

"The second wave of deindus-trialization begins to hit Merthyr Tydfil, and that's the start of pit closures just before the Second World War. People begin to be very worried about this, so they start bringing in new factories, new businesses. Manufacturing started to replace coal. We got a huge Hoover washing machine plant. At its peak, it employed five thousand people.

"What I want you to notice is Merthyr Tydfil has had three different periods of deindus-trialization, different changes in the industry, and each time the wave has passed over, but of course Merthyr Tydfil is still there, south Wales is still there, and I think that's really import-ant to remember. What you need to know is that the place survived. How it has survived is a larger question."

The emblem of the South Wales Miners' Federation.

on October 24, 1898. This regional union, commonly known as "the Fed," eventually played a major role in shaping trade union policy in the UK (Francis and Smith 1998, xvi).

As the miners' union gained members and political strength, their leaders increased demands for a safer workplace and fair pay. At times, conflicts between labor and management became heated. In 1910, the Cambrian Combine strike exploded into the Tonypandy riots that rocked south Wales (Curtis 2013). Then, in 1926, miners throughout Great Britain, including south Wales, led a general strike of all industrial workers, which to this day remains the only nation-wide strike in UK history (Gildart 2011). In the aftermath of the strike, south Wales miners led "hunger marches" to London to highlight the difficult conditions in the coalfields (Curtis 2013).

Parc and Dare Miners' Institute in Treorchy, Wales.

During these struggles, the South Wales Miners' Federation became deeply embedded in community life in the valleys. The union's support of education and cultural activities extended the benefits of the union beyond the mine site, or colliery grounds, as they are referred to in the UK. Miners donated a portion of their weekly wages to support the building of miners' institutes (also called workingmen's institutes or welfare halls) in their communities (Davies and Jenkins 2008). These buildings provided local libraries, served as performance venues, and became an important communal space. A broad spectrum of community members gathered in these halls to watch movies, hear opera, take continuing education courses, or sing together. Many present-day coalfield residents informed me that the

Author Tom Hansell (right) talks with Dai Charles Davies (left) in the Onllwyn Miners' Welfare hall.

union's commitment to offering miners educational opportunities was a key factor in building strength and resilience in the coalfield communities of south Wales.

Sitting in the Onllwyn Miners' Welfare Hall, retired miner Dai Charles Davies explained how this hall was created, and reflected on his career:

> I was born in 1924. I was born in Pantyffordd, then we left there when I was four years of age and went to live in this place here [Onllwyn]. I've been here ever since.
>
> Originally my father worked for McTurk in Cray, at the Cray waterworks. That was a big waterworks up there near Swansea. He worked with McTurk for some years and then he went to the colliery in Abercraf. International Colliery, it was called then. There were people from Spain, Portugal, and different countries that worked there. He worked there.
>
> When he worked for McTurk we rented a house for ten bob a week. My mother kept chickens and eggs and took them to the market to sell. He realized that there was better

money in the colliery so he went to the international colliery in Abercraf. Then down to Craig y Nos. Then he left that colliery and went to work in Onllwyn.

I worked in Onllwyn Colliery first. I spent a year there. My brother was working in Banwen, the colliery at the top of the valley. He persuaded me to come up there and work with him. And then, when that closed they sent us to Cefyn Coed, the deep mine in Crynant. I worked there for a period and when that closed I went to Treforgan and that's where I finished mining.

I was a member of the miners' welfare committee in 1946 and we didn't have tennis courts or anything then. As a consequence of Dai Francis being elected to the South Wales Miners' Federation, he developed the idea that we should raise money to show the people at the center in Cardiff that we were serious about the idea of having a hall—a welfare hall. We formed different committees to raise money. We had jazz bands and different things. I think we raised about two thousand pounds. That indicated to the people in Cardiff that we should have a welfare hall.

Before this welfare hall was here, we didn't have a cinema. People had to travel down to Seven Sisters, three mines down, for a cinema. The idea of a welfare hall caught on with the public. We were subscribed to by the collieries, Onllwyn and Banwen collieries and the washery, they provided the income for the hall. It caught the imagination of the people and they supported everything. Of course it was 1955 when the hall was erected. We had money from the collieries then to pay for this. Once that dried up, Dai Francis had the idea of a bar. Now that and the choir is what keeps us going.

We are concerned about keeping the choir going. We need to recruit young people, and we have failed to do that so far, we're a bloody old man's choir in fact! (Laughs.) But the effort of the people over the years has been tremendous, keeping this miners' hall open is all voluntary work, nobody gets paid.

NATIONALIZATION

After World War II, organized labor was at the height of its political power. The Labour Party took control of British government during a landslide election in 1945, when Clement Attlee became prime minister (Brown 2012). Faced with the daunting task of rebuilding infrastructure that had been devastated during the war, Labour's solution was to nationalize vital industries, including the coal industry, using public funds to coordinate reconstruction efforts.

The period of nationalization was recent enough that I was able to access first-person accounts. One of my best sources of information came from an interview that *After Coal* advisors Helen Lewis and John Gaventa recorded with Dai Francis (the father of historian and project advisor Hywel Francis) to commemorate his retirement as general secretary of the South Wales Area of the National Union of Mineworkers in 1976.

During the interview, Dai Francis provided an eyewitness account of the nationalization of his home colliery in Onllwyn, Wales:

> This was a great day. What we called Vesting Day [January 1, 1947]. The mining industry of Britain was to be vested in the state. I was scheduled to speak at other collieries, but I wanted to be at home. The speech I wrote was to be delivered in my own village, in the colliery where I had always worked. And it was a great moment. I remember watching the hoisting of the flag and feeling proud that this little colliery where I had worked was no longer in the hands of the coal owners (Lewis 1976a).

National Coal Board sign.

It is worth highlighting how differently Britain's nationalized coal industry operated from the privately owned coal industry in the US. In the UK, the National Coal Board was established under the Coal Industry Nationalisation Act of 1946 as a branch of the central government to control production, employment, and safety regulations. While production was, of course, important, the coal board also considered the well-being of the communities where mines were located.

Dai Francis was quick to point out the difference:

> The mines in the States are being run for what purpose? Not to produce the energy that the US requires. That's of secondary importance. The most important thing as far as those coal owners in America are concerned is profit. The emphasis is on profit.
>
> We believe in production for use, not for profit. That's what socialism is. Here in Britain there are many who say, "We've got to make a profit, we've got to make the industry pay." But I believe that approach to mining has got to be removed . . . We need greater worker's control . . . The product was nationalized, but the distribution was left in the arms of private enterprise. The whole of the coal-mining industry should have been nationalized. That was one big mistake, that was (Lewis 1976a).

COAL BOARD CONTROVERSY

Although miners gained power through the process of nationalization, tensions remained between union members and the National Coal Board. The miners I spoke with agreed that while safety increased significantly under nationalization, coal employment eventually decreased, with devastating effects on many mining villages.

In order to compete with oil during the 1960s, the National Coal Board developed a policy of investing in what were known as "super pits." This meant closing small collieries in the center of mining villages in order to concentrate resources into large consolidated collieries (Curtis 2013). Miners were expected to leave the pit in their community to work in these large modern mines. However, many men resisted this change and abandoned their careers in mining in order to commute to factory jobs in population centers. This loss in manpower complicated the coal board's task of managing these pits and resulted in an accelerating cycle of decline.

An article in a union newsletter summarized the concerns of coal communities: "One of the main reasons for men leaving the industry at this rate is the feeling of insecurity that has been engendered in the industry by the Coal Board with its policies of closing pits" (Mineworkers 1965). Clearly, the 1960s were the beginning of a long-term decline in coal for the south Wales valleys.

I interviewed two miners who had weathered this storm. Terry Thomas and Eric Davies both started their careers at the Garngoch number three colliery in Gorseinon, Wales. When Garngoch was closed in 1965, they were transferred to the super pit at Brynlliw. In 2013, I interviewed them at the site of the Brynlliw Colliery, now an empty field.

> **ERIC DAVIES:** I started here [in] September 1965. We had been working at Garngoch number three colliery, but they needed men here. People were leaving the pits, so what they decided to do was to close a local colliery, send the five hundred odd men here and that was it. In fact, if you look at the men that came here from close collieries, it was something like fourteen different collieries that closed and their men came into this pit. But then of course eventually this pit went as well.

> **TERRY THOMAS:** I came from Garngoch along with Eric in 1965. In the early sixties there were caravans on the site along here, trailers for the miners that came down from Durham to work in the mines in south Wales. They were living in the trailers right on the surface of this colliery.

ENERGY CRISIS AND EXCHANGE

The oil embargo of 1973 set off the energy crisis that helped define the 1970s. Demand for coal skyrocketed, prices rose, and new life was breathed into a declining industry. In 1974, the National Union of Mineworkers took advantage of the increased dependence on coal and called for a national strike to increase wages and benefits. Tory prime minister Edward Heath refused to give in to the workers' demands, and responded to the rolling power outages and disruption caused by the strike by calling a national election based on the slogan "Who Governs Britain?" (Hansell 2016).

In the midst of the national strike, an American student named John Gaventa arrived in the Welsh coalfields with early portable video equipment. He enlisted the technical help of a young Welshman named Richard Greatrex and received assistance from historian Hywel Francis at Swansea University. They started documenting the national miners' strike in Wales with the idea of sharing these stories with striking miners in Harlan County, Kentucky. A 1974 article by John Gaventa published in *Mountain Review* explains the intent behind this exchange (see sidebar on opposite page). Appalachian sociologist Helen Lewis joined the project when she moved to Wales in 1975. She eventually donated the videotapes made by this group to the W.L. Eury Appalachian Collection at Appalachian State University, where they provided inspiration for the After Coal project.

When I interviewed Welsh miner and labor leader Terry Thomas, I asked him about a miners' march to Sophia Gardens in Cardiff that Gaventa and Greatrex had captured on video during the 1974 strike. He told me: "The streets were lined with people who were applauding us as we were marching. It was an unbelievable feeling, a feeling that told us what we were doing was right. We knew at that time that we were going to win that strike, and that was a wonderful feeling."

...I'm sure everyone in this room was shocked by the situation handed to us on videotape...they're back in the situation we were twenty-six years ago. . .when they're up against small owners--which we had all over our coalfields around here - - owning our homes, owning our means of production, owning our food, owning our furniture shops...we are in a struggle ourselves. Although we are one hundred percent NUM or union membership, although we are nationalized, it is still necessary for us in 1974 to have a national strike so we can gain or improve our standard of living.

...I'll tell you one thing, if we had to face things like that--guns and violence and jail--I don't know if we'd be out of the picket line...

VIDEO AND MINERS:
APPALACHIA AND WALES

by John Gaventa

These were but some of the comments made by Welsh miners in Britain after watching a videotape of striking miners at the Brookside mines in Harlan County, Kentucky. The time was February 1974, and the Welsh miners themselves were on strike, in an effort to gain an increase in wages--then less than half those of most Kentucky miners. But what impressed these Welsh miners was not the wages of the American miners, but the conditions they were up against--poor safety, anti-unionism, company thugs, "kangaroo" courts. And the determination of the men, women and children of Harlan in their struggle to win basic union rights reminded these Welshmen of their own similar struggles years before.

In 1934 a Welshman visiting Appalachia wrote about the coal camp conditions, "The miners of West Virginia are paid higher than the miners of South Wales but the miners of South Wales would not change places with the miners of West Virginia, because the South Wales miners have won a measure of freedom from industrial autocracy and of political independence which would be denied them in West Virginia." Some forty years later, as these men watched the videotape of the miners up against Duke Power Company in Harlan, one sensed that the same theme would still hold true.

This was only one impression of many that grew from a project that I've been carrying out, away for a while from Appalachia, studying and researching in England. The idea of the project was simple: to use a videotape portapack to allow Appalachians to communicate with and learn from people in similar situations--people who were involved in or had already faced similar battles to those we face in the mountains at home. This particular project was on working conditions, union organization, and union democracy among the British miners. Last November I had taken some videotape in Harlan, including interviews with

union organizers--Mike Trobovich and others--speaking at that memorable rally in Evarts, and women from the picket line describing their harassment by scabs and the "justice" they received from an "operator-judge." Now, back in Britain, the miners' strike provided an opportunity to show that tape and to spark some dealogue on the contrasting conditions of British and American coal mining communities.

We had learned in projects in east Tennessee that videotape can help mountain people overcome their isolation and communicate directly with each other from hollow to hollow. And we've learned, too, that that direct, alternative communication is essential because the commercial media--controlled by corporate interests and aimed at middle class audiences--is rarely going to tell the story that must be told of the Appalachian people.

The lessons from the mountains easily apply to the way the media views and treats working people in similar situations outside of the mountains. For instance, in Britain the miners felt that the television was not giving them a fair representation--"It is just a public relations exercise with the television and the mass media when they say the public is against us," said Terry Thomas, a miner at the Brynlliw Colliery. In the United States, the media portrayed the striking British miners as power-hungry radicals, threatening to bring the country to its knees, while little attention was drawn to the facts that they were the lowest paid miners in all of Europe and that even the public opinion polls showed them to have the support of a majority of the British public.

One correspondent from *Time* magazine ventured into the South Wales valleys and came out headlining his article with "At the End of the World in Wales," adding "it would seem strange to any outsider that generations of valley life

should prefer to live in the midst of such cruel memories." He might, one suspects, have written the same about Kentucky, but the statement shows only too clearly a lack of understanding for the quality of life that is shared by mining communities, whether Appalachian or Welsh. Perhaps that quality can only be appreciated--and communicated--by those who share it.

In those all-too-rare moments when the commercial media does capture effectively the struggles and strengths of such mining communities, the effects in other similar communities demonstrate the impact such communication can have. People in a strip-mined valley in Wales still recalled vividly and with some dismay *The Stripping of Appalachia*, the powerful film shown over a year before on the problems brought upon the people of the Clear Fork Valley by the policies of a British-owned land company, American Association. Made by a British film crew with the participation of local residents, the documentary has also been circulated throughout the Appalachias and elsewhere in the United States as an effective communications tool for communities there which share the same problems.

Similarly, during the strike in Britain a film was shown by the British Broadcasting Corporation on the movement for union reform, problems of safety and the callousness of the corporations in Appalachia. That film, too, provoked strong reactions from the miners in the Welsh valleys. Often the responses were ones of shock, particularly to safety conditions and lack of enforcement of safety regulations in the Appalachian mines. "I personally would sooner starve than work in the conditions of their support roofs," said Ken Williams, a face worker at the Treforgen Colliery. And he added, "In this country, you can be sent out of the pit if an official comes and you are working in exposed dirt of over a yard. And if a section of workmen are working in unsafe conditions we have a right to go to the lodge (local union) and ask the lodge for a special inspection. They grant it–without argument." For the communication process to have been complete, reactions like these should have been taken back to the Appalachias–but that might be *too* much to ask from the corporate media.

Such instances of communication between mining communities of different countries reinforce the points we already have learned in Appalachia. We cannot depend upon the outside commercial media; there is a need for an alternative media, controlled by and provided for our own communities; and such media, when used effectively, can be a strong force in the development of consciousness, organization, and dialogue within our own communities and between other communities.

It was with these ideas in mind, anyway, that I took a video portapack to the valleys of South Wales during the British coal miners' strike of 1974. I took some thirty reels of tape in union halls, working men's clubs, homes, and mine canteens. They were of meetings, interviews, reactions to American tapes, singing and rallies. There were the usual frustrating technical problems, complicated by differences in British electricity and by incompatibilitie of British substitutes when my gear blew out. But one hour-long edited program, *Strike '74*, is now back in Appalachia, and other programs may be produced from the material soon.

The tape itself brings out what was learned about the differences between the Appalachian and Welsh coalfields. But also of interest was the *video process* that was at work. From the beginning it was distinguished from that of the normal media: the tape was not to be shown publicly in Britain until after the strike; the audience was American miners, with whom these miners felt some solidarity; the participants would have a chance to view rough material, advise on editing, and see the finished version. A process was thus at work in which they keenly participated and in which they were anxious to share. Soon the miners were telling me what they wanted taped and allowing me in meetings from which the commercial media were excluded–always after a democratic vote.

A few comments should be made about three particularly interesting aspects of this process.

Video and Feedback. Whenever possible, the videotapes–be they interviews, meetings, choir practices, rugby matches–were shown back to the participants. The feedback sparked responses of further conversations and reflections of pride or of self-criticism. Everyone enjoys viewing themselves, but the impact is something special upon people who are normally denied the chance to watch themselves–or anybody like themselves–favorably or realistically portrayed on the television screen.

One particularly dramatic example of this feedback process occurred one night early in the strike in a miners' club in the town of Clydach. I did a series of interviews about the events leading up to the strike. "It all started in 1926," said one long-time union man, and he continued by describing in detail for thirty minutes

the history of the South Wales miners since. The video was then taken into the main room, set on a table, and played back in its rough form. A friend wrote down for me what he observed happening:

"It is difficult to account for the sudden grip the video screen had on all the people there drinking beer, listening to music, talking or reading. The clutch of the set was immediate and real.

No one moved to turn off this playback as they did a BBC program in a miners' lodge up the road. Virtually no one in the room discussed any other topic.

An unusually long attention span seemed to prevail.

The video set almost emanated an air of intervention: something had changed; someone new or something new had happened which demanded a response."

The response took the form of discussion about the issues raised, the people, the community. The tape seemed to provide a reason to recall memories, and to view the current situation afresh. "Evan's been at it a long time, you know..." said one viewer, and then went on to describe the deterioration of his mining community, with the pits closing.

But, even more, the tape seemed to intensify a sense of unity and pride among the viewers themselves, most of whom were striking miners and their families. When the club closed down that night, something had happened. As my friend described it, "This feedback effect touched those actually on the screen much more than the others, but most people seemed to share a sense of identity heightened by seeing their own people on the screen and their own situation, or a situation so similar as to invite personal comparison. This event had such a tonic effect that it can hardly be viewed on the same level as TV in general. In fact, the videotape made it an event.

Video and Organizing. Two features more than any others stand out about the miners and their organization in Britain. The first is their internal union democracy; officials are elected, members participate in making decisions and formulating policies; leaders are held responsible for how they represent the men. The second impressive feature is the solidarity among the miners and with other workers. When the men went on strike, virtually all coal production in the country came to a stop. And, more effectively, the truckers, railwaymen, and dockers refused to transport any of the stockpiled coal, while many workmen of other industries refused to use it until the strike was settled. Such organization is difficult to attain or maintain in any situation. In a small way, we found that video could help.

When leaders of the Brynlliw mine went to meet with the leaders of a local steelworkers' union, they asked us to come along and record the meeting. The tape of their presentation could then be used for the miners back in the local union to see how they were being represented by their leaders. In fact, some fifty of the men spent an entire Sunday afternoon watching the two hours of tape, criticizing or applauding their spokesmen, and thinking of how the presentation could be even more effective

the next time.

Video and Exchange. While the video thus helped in the organizing situation, perhaps of most interest was the sharing of video tapes from other miners, in this case, striking miners at Brookside. The tapes were shown at several sessions in local miners' clubs or centers to groups of about forty. The effects of the showings were profound; the miners watched with interest, sympathy, and shock. They then were asked what they would have to say to the American miner, particularly in view of the new leadership of the United Mine Workers, the conditions faced in places like Harlan, and the current negotiations for renewal of the contract which expires in November. The responses were videotaped. Here are a few of them:

...You've seen this before. Fifty years ago it was here. The vested interests of the judge in his decisions, the scab unions, all that kind of thing.

...We have been through that. And another good thing that I noticed here as during that time, the women were very strong with us, and it was excellent then when the women were there to help.

...Something similar but not as far as the fellow threatening with the gun, you know. We had them with their batons, knocking us on the heads, and that kind of thing, but not to the extent of anybody being over saying "Shoot him."

...Well, John L. Lewis, of course his parents came from Pontardulaism, not very far from here. And as you say, John L. became a very popular hero in this country, with the miners themselves, (but) whilst we have to recognize that John L. was making a very powerful union at the start they were regarding one man as (the spokesman) and not the trade union.

...Once a man was an agent he was the boss...But with the movement that sprang up in South Wales in the 1930s...we said we've got to change this, the rank and file must decide the policy of the union and not the leaders as it had been up to then you see. So from there on we had the rank and file leadership, and that is where the grass roots were started, and that is what has to commence in the States again. It shows now that there is a change in policy and that is where the strength will be.

...It is a struggle between the haves and have-nots you know. There is a strong feeling wherever you go, there is a close liaison between the separate sections of the working class now.

...I think the biggest thing...is that we are one hundred percent organized, unless you have got that, that is the foundation...and that is the biggest lesson that we can quote to America, that we are organized one hundred percent.

...I believe that the point must be made clear, not for them to think that in Wales we are born with this ready made organization–it is not, it is a gradual process.

The Welsh miners had many such comments about America. But in the process of exchange they, too, had gained an appreciation of another community, somewhat similar, yet different from their own. One miner concluded one session by saying, "Not only are you go-

VIDEO

ing to take ideas back from here now, from the South Wales miners or from the British miners to the United Mine-workers of America, you have certainly taught us something, and I am sure we have all learned something tonight." Or, in another meeting, "an exchange of ideas has happened...If ideas hadn't come into this isolated valley before, where would we be now?" Yet another miner added, "We were skeptical about this at first, but now we think that our union ought to have more things like this for us. It was educational, but it wasn't a lecture. These guys wouldn't have put up with a lecture." And then their conversations moved back to new thinking about their present struggles.

I do not know what response the tapes will bring in America. But no matter what the outcome there, the process itself has established a beginning bond between mining communities to whom the opportunity to share experiences is usually denied. Hopefully, that bond will have the chance to grow through further exchanges sponsored by the respective unions, by foundations, or by other groups. But, again, the words of South Wales miners say what must be said about this particular effort. After seeing an edited version to which I had requested criticisms, Doug Thomas of Gwmturch expressed his appreciation for the "contribution of coming from the States and showing what's happening in the States and in return showing what's happening in Wales. That to me is the important point of all these films—not to contradict what we've seen or to say what's missing, but the message the film states—the way to achieve unity inside our mining communities...With all due respects, we hope that the small contribution of the videotape going back will help our American friends."

Of course, it doesn't take another country for this communication process to work. There is something in the use of video for feedback, organizing, and exchange that can be of value to workers, poor people, minorities—groups anywhere that are denied the right to speak to each other or be heard by others through the corporate media. The growing movement for such alternative communication—of which this project and this new journal, the *Mountain Review*, are an indication—must continue throughout dispossessed communities in Appalachia and elsewhere.

John Gaventa, a Rhodes Scholar in Political Science, has conducted a tax evaluation of East Tennessee Coal companies for the Vanderbilt Student Health Coalition.

Prime Minister Heath was soundly defeated in the national election held during the strike and the new Labour government quickly settled the strike on favorable terms with the miners. In the videotapes recorded during that time period, I discovered that Gaventa had pointed the camera at a television set during the BBC's nightly news. The video shows a well-coiffed newscaster reporting: "An end to the miners' strike: Their leaders voted tonight to recommend acceptance of a coal board one-million-pound pay package. The revised package, guaranteeing miners between thirty-two and forty-five pounds per week basic pay, is double the original offer and is fully approved by the government" (Lewis 1976b).

The increased demand for coal during the 1970s provided a ray of hope for the Welsh coalfields. With Helen Lewis's help, John Gaventa and Richard Greatrex expanded their videotape project from an exclusive focus on coal miners to include scenes of vibrant community life throughout the south Wales valleys. The

trio named this project "The Welsh Tapes" and this body of work shaped their future careers.[1]

For me, "The Welsh Tapes" provided a crash course in Welsh coalfield history. The wavy black-and-white images provided a window

ORAL HISTORIES FROM TERRY THOMAS AND ERIC DAVIES

PAT BEAVER: Tell us about the first time going underground.

ERIC DAVIES: I went to work with a chap that Terry had worked with before. A chap by the name of Will Thomas, or Will Shilling, that was his nickname. Now because I was fifteen years of age and to be supervised closely underground, I went into the coal seam with Will Thomas.

TERRY THOMAS: Now I had been working with Will Thomas (we called him Will Shilling), but because I was a few years older, the training regulations allowed me to be able to go and work on what was partially a training face, but partially a production face. So what it meant was that I had to get thrown out [of] my nice comfortable place to make room for this guy (gestures to Eric) back in 1960 in Garngoch number three colliery (laughs). I had to go to a new coalface where we

were doing some serious mining so this guy could have this comfortable working place with Will Shilling.

The funny thing is, especially in south Wales, most people have nicknames. And very often you'd know the person by the nickname and you wouldn't know their proper name. For years and years there was a manager we had whose nickname was Sooty. I can remember calling him Mr. Sooty because I had no idea what his real name was!

TOM HANSELL: What were your nicknames?

ERIC DAVIES: He was known as the Chin.

TERRY THOMAS: Maybe because of my features. And I can tell you that at the time, this one (pointing to Eric) was either Citizen Smith or Captain Pugwash.

ERIC DAVIES: And Chairman Mao!

into the life and culture of the south Wales coalfields during the mid-seventies. I spent hours poring over unedited footage of underground mining, in which the Welsh miners were keen to explain every safety precaution in great detail. I watched miners' choirs create complex harmonies from the plain words of Welsh hymns and African American spirituals. Schoolchildren explained the significance of pinning leeks to their blouses on St. David's Day, and pub patrons spontaneously recited poetry or burst into song.

Reflecting on the experience, Helen Lewis told me, "I think what we did with these films is really document . . . the last generation of those communities as real mining communities."

ORAL HISTORIES FROM TERRY THOMAS AND ERIC DAVIES

TERRY THOMAS: Yes, he had several nicknames.

ERIC DAVIES: And Eric the Red, of course (laughs).

TERRY THOMAS: There were some happy times as well as the hard and bitter times, too. When we arrived here, there was just under one thousand men working here underground. Looking at this now it is difficult to think of where everything was. It is now thirty years since I last stood on this site of where Brynlliw Colliery was. That's why it is difficult to imagine the site now. You wouldn't even think that there had been a coal mine on this site looking at it now. When there were once a thousand coal miners working underground at this colliery.

TOM HANSELL: Do you remember your first day here?

ERIC DAVIES: Yes, it was September 1965. I'd just come back from holidays. I came here, picked my tools up—miners' tools up—and took them to the shaft, down the lift, and after that we went into the place of work. It was in the west and it was called the S7 coalface.

PAT BEAVER: What were your duties?

ERIC DAVIES: I was a face worker—working on the coalface where the coal was being cut. I was supporting the roof and doing whatever was needed on the coalface.

TERRY THOMAS: I remember when I first went to Brynlliw, I was put to work on installing a new coalface. Really my work was not much different here than it was at Garngoch. I was working on the coalface taking care of the mechanical equipment that was cutting the coal; that was my first memories of here.

APPALACHIA

Coal came in the twentieth century like blazing guns. And that was a good thing back then, but now, here it is, the twenty-first century, and coal is a thing of the past.

—Carl Shoupe, retired miner, Harlan County, Kentucky

INDUSTRIALIZATION

In contrast to Wales, where the coal industry took root in the early nineteenth century, development of the Appalachian coalfields began in the late nineteenth century and grew with lightning speed.

ORAL HISTORIES FROM TERRY THOMAS AND ERIC DAVIES

But, the important thing was that the miners working in Garngoch were working very low seams, down to two foot three inches. Whereas, when we came here I was working in a six-foot seam where you were standing upright. The environment was completely different, in the seams of coal we were working in Brynlliw as opposed [to] the seams of coal we were working at the mine where we came from. Of course it was a much, much bigger operation.

TOM HANSELL: Do you remember your last day here?

ERIC DAVIES: Yes, it was the twenty-fourth of December, 1983. That's the last day when I went out of this colliery. Actually, I think I was the last person to leave here other than the watchman. He was still here for some years afterwards.

TOM HANSELL: What did that feel like to be the last person?

ERIC DAVIES: It was a funny feeling really. We went out of the gate, jumped in the car, and away we go. Just look back, jump in the car, and off you went. The rest is history I suppose.

We've gone on to do other things and our people have diversified. Many of our friends and colleagues are going through all sorts of different jobs. Some went to other pits, many of them did, and of course in a few years those pits closed as well. Some were old enough to have pensions, but many went to other jobs and it's amazing what jobs miners went and done: school caretakers, businessmen. It's funny how people diversify. Many did find work but no major employer came in its place. And many were not particularly well-paid jobs either.

These ancient mountains held all the necessary ingredients for the industrial revolution: coal, timber, iron, and water. Following the American Civil War, the Appalachian Mountains were promoted as a magnificent new investment opportunity for capitalists. As a result, these coalfields experienced monumental growth in a short time period. Historian Ron Eller explains: "By 1900, coal production in the region had tripled, and in the next three decades it multiplied again more than fivefold, coming to account for almost 80 percent of national production" (1982, 128).

The eastern industrialists who financed Appalachia's railroads and coal mines successfully concentrated land ownership and mineral rights. By 1910, eighty-five percent of the mineral rights in eastern Kentucky were owned by nonresidents (Caudill 1963). This fast-paced

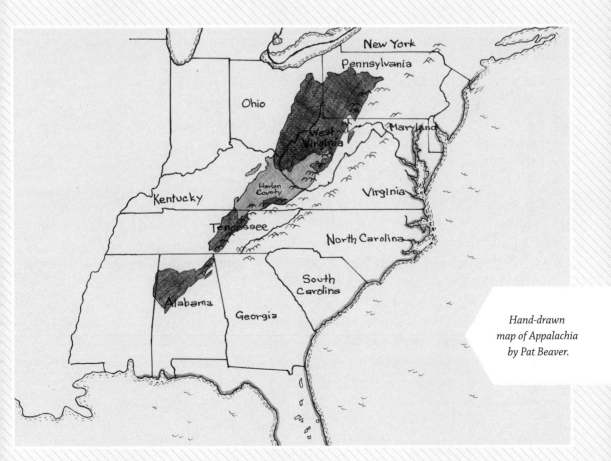

Hand-drawn map of Appalachia by Pat Beaver.

Construction of the power plant in Lynch, Kentucky.

NOV 1 19

TIMELINE BY ANGELA WILEY

	1750–1860	1831	1842	1880–1920	1890	1898
APPALACHIA				Growth of the timber and coal industries brings immigrants and rapid growth to the region.	The United Mine Workers of America (UMWA) is formed.	
SOUTH WALES	Growth of the iron and coal industries brings immigrants and rapid growth to the region.	The Merthyr Rising is the first major labor unrest in Wales and the military is called in to restore order.	The Mines and Collieries Act prohibits underground work for women and girls, and for boys under the age of ten.			The South Wales Miners' Federation (aka "The Fed") emerges as the leading miners' union in south Wales.

change in land ownership into the hands of big business provided America with access to the resources required to develop into a world power. However, control of vast quantities of land allowed absentee corporations to block economic diversification and dramatically limited opportunities for the people of Appalachia over the next century.

The photographs on pages 45 and 46 from local supervisors' reports to the US Coal and Coke Company (later to become US Steel) illustrate the planning and construction of a model coal town to be built in Harlan County, Kentucky, and named after the company president, Thomas Lynch. The company started building in 1917 and the mine opened in 1921.

COMPARING THE HISTORIES OF CENTRAL APPALACHIA AND SOUTH WALES

1907	1910	1912–13	1914–18	1915	1920
The Monongah Mine disaster kills at least 361 miners in West Virginia.		Paint Creek–Cabin Creek strike in West Virginia. Miners list demands that include the right to organize into a union and fair weighing of coal.	World War I: Coal is essential to the war effort. As a result, the government guarantees high wages for coal miners.		The Matewan Massacre in West Virginia leaves 11 people dead in a clash between mine guards and miners.
	The Cambrian Combine strike explodes into the Tonypandy riots as miners clash with mine owners and police over wages.	Senghenydd Colliery disaster kills 439 miners in Glamorgan.	World War I: Coal is essential to the war effort. As a result, the government guarantees high wages for coal miners.	Coal industry reaches peak employment of 250,000 in Wales.	

MIGRATION

The early twentieth-century coal boom brought dramatic population increases to the Appalachian coalfields through the recruitment of a diverse labor force. Locals moved from farms to the new coal towns for the promise of work for wages. African Americans moved to the coalfields to escape the oppressive sharecropping system in the Deep South. American workers were joined by an international labor force from Italy, Hungary, Greece, Poland, and other European nations—including Welsh, English, and Irish immigrants who came to the southern mountains by way of Pennsylvania. Residents of Lynch, Kentucky, recall that about forty different languages were spoken in town during the 1920s, prompting sociologist Bill Turner to call his hometown "a little New York" (Barret 1987).

COMPARING THE HISTORIES OF CENTRAL APPALACHIA AND SOUTH WALES

	1921	1926	1930	1935	1939
APPALACHIA	The Battle of Blair Mountain becomes the largest armed insurrection in the United States since the Civil War. Approximately 10,000 armed miners marched in southern West Virginia in an attempt to unionize the region. After several days of fighting with mine guards, the National Guard was called in to disperse the miners.		Central Appalachia employs 160,000 in the coal industry.	The National Labor Relations Act is passed, guaranteeing workers the rights to form a union and collectively bargain.	
SOUTH WALES		Miners lead a national general strike throughout the UK to protest wage cuts. In the aftermath of the strike, south Wales miners lead "hunger marches" to London.			Great Britain enters World War II. The demand for coal increases.

Image of the construction of Lynch, Kentucky.

=JuNE=1=1919=

COMPARING THE HISTORIES OF CENTRAL APPALACHIA AND SOUTH WALES

1941	1947	1950	1960	1966	1968
The United States enters World War II. The demand for coal increases.	The Taft-Hartley Act amends the National Labor Relations Act. Organized labor faces new restrictions on union activities.	Central Appalachia reaches peak employment of 196,000 in the coal industry.	Mechanization of the coal industry results in job losses and out-migration; more than one million residents leave Appalachia between 1950 and 1960.		The Farmington Mine disaster in West Virginia kills 78 miners in an explosion.
	The National Coal Board is created to run a newly nationalized coal industry in Wales.	South Wales employs 102,000 in the coal industry.	South Wales employs 84,000 in the coal industry.	The Aberfan disaster. A coal tip collapses, killing 116 children and 28 adults.	

COMPARING THE HISTORIES OF CENTRAL APPALACHIA AND SOUTH WALES

	APPALACHIA				
	The Federal Coal Mine Health and Safety Act requires coal mines to enact stronger preventative health and safety measures.	Central Appalachia employs 38,000 in the coal industry.	A slurry dam bursts, causing the Buffalo Creek Flood, which leaves 118 dead and 4,000 homeless.	Arab members of the Organization of the Petroleum Exporting Countries declare an oil embargo, limiting exports to the US and raising oil prices worldwide. Coal prices also increase, causing a boom in Appalachia.	Strip mining increases in Appalachia as a result of the energy crisis. The Brookside strike starts in Harlan County, Kentucky, as miners at Duke Energy's Eastover mine vote for union representation.
	1969	**1970**	**1972**		**1973**
SOUTH WALES		South Wales employs 61,000 in the coal industry.	A national miners' strike results in increased wages for Welsh miners.	Arab members of the Organization of the Petroleum Exporting Countries declare an oil embargo, limiting exports to the US and raising oil prices worldwide. Coal prices also increase, causing a boom in Wales.	Hywel Francis establishes the South Wales Miners' Library at Swansea University.

When US Steel opened a mine here, they hired a guy they called Limehouse to go and get the black people in Alabama and bring them to Kentucky by night. The reason they had to do it by night was because it was illegal, they were leaving sharecropping and the plantation owners did not want them to go. That's how most of the blacks who worked in the coal mine got here—most of them came from Alabama.

—Rutland Melton, native of Lynch, Kentucky

Appalachian coal communities were segregated by race and ethnicity. Historian Ronald Lewis uncovered correspondence in which coal executives in Philadelphia encouraged their managers in the coalfields to hire a "judicious mixture" of Appalachians, African Americans, and European immigrants in order to discourage union organizing (Lewis 2009). During this era, white and black miners had their own

COMPARING THE HISTORIES OF CENTRAL APPALACHIA AND SOUTH WALES

1974	1975	1977	1979	1980
The Brookside strike is settled with miners winning union representation.		The Surface Mining Control and Reclamation Act sets forth standards for regulating the environmental effects of coal mining in the United States. A loophole allows mountaintop-removal mining.		Central Appalachia employs 97,000 in the coal industry.
A national miners' strike in the UK results in a new Labour government. John Gaventa receives a Rhodes scholarship to study at Oxford. He meets Richard Greatrex and begins to document the national miners' strike in Wales.	Helen Lewis receives a grant from the National Science Foundation and moves to Wales. She eventually develops an exchange between Appalachia and Wales that seeds the After Coal project.		Welsh miners visit the United States, accompanied by Hywel Francis from Swansea University. Helen Lewis from the Highlander Center and Pat Beaver from Appalachian State University help host their visit.	South Wales employs 20,000 in the coal industry. Global competition decreases demand for Welsh coal.

segregated washhouses and their children attended separate schools. This racial segregation helped coal companies maintain a high level of social control in the towns they built and owned.

ORGANIZED LABOR

During the early decades of the twentieth century coal companies tightly controlled miners' lives. Workers were often paid in company scrip, redeemable only at the company store. They lived in company-owned coal camps and coal towns, where private corporations owned the houses they lived in, the schools that educated their children, and the churches where they worshipped. Private detectives and mine guards enforced company policy, and violations meant not just the loss of a job but eviction from their home.

COMPARING THE HISTORIES OF CENTRAL APPALACHIA AND SOUTH WALES

APPALACHIA	1981	1984	1988	1990	1997	2000
	The United Mine Workers of America selectively strike against the A.T. Massey Coal Company.	Kentucky voters change the state's constitution to outlaw the broad form deed, which allowed for strip mining without a landowner's permission.		Central Appalachia employs 48,000 in the coal industry.	Pat Beaver, Helen Lewis, and John Gaventa visit Wales for a reunion with the participants in the 1979 miners' exchange.	Central Appalachia employs 29,000 in the coal industry. In Martin County, Kentucky, 300 million gallons of coal slurry spill into tributaries of the Big Sandy River.
SOUTH WALES	Pat Beaver visits Wales, continuing the cultural exchange between the coalfield communities of south Wales and Appalachia.	A yearlong national miners' strike signals the end of coal for south Wales.		South Wales employs 1,000 in the coal industry.	Hywel Francis and Mair Francis visit Appalachian State University and the Highlander Center in Tennessee.	South Wales employs 500 in the coal industry. UNESCO designates the Big Pit National Coal Museum a World Heritage Site.

To improve their situation, miners organized into unions. Many of their early demands were designed to increase miners' control over their lives. Miners' demands included:

- the right to organize;

- recognition of their constitutional rights to free speech and assembly;

- alternatives to company stores;

- an end to the practice of using mine guards;

- installation of scales at all mines for accurately weighing coal;

- that unions be allowed to hire their own checkweighmen to make sure the companies were not cheating the miners (Corbin 1981).

COMPARING THE HISTORIES OF CENTRAL APPALACHIA AND SOUTH WALES

2001	2006	2007	2008	2010	2014
	Onllwyn Choir visits Appalachian State University.	First year with no underground coal-mining fatalities in Kentucky since 1890.	TVA's Kingston Fossil Plant spills 1.1 billion gallons of coal ash into nearby rivers.	Upper Big Branch Mine disaster kills 29 miners in West Virginia. The tragedy opens a federal investigation into safety practices.	Central Appalachia employs 18,000 in the coal industry.
Pat Beaver and Helen Lewis teach an Appalachian State University study abroad course in Wales titled "South Wales in a Post-Coal Society." The course continues to be offered today.					South Wales employs 800 in the coal industry.

These photos document the segregated school system in the neighboring coal camps of Benham and Lynch, Kentucky.

PERSONAL HISTORY: RUTLAND MELTON

I'm a retired coal miner, I worked in the coal mines for twenty-three years, when I worked in the mine it was a union mine. In 1998 they shut the mine down and they kept it shut down for a year. But when they reopened it, they reopened it nonunion. Anybody who had been a union miner worker, they denied them. So I retired early at the age of forty-nine.

My grandfather was named McKinley Melton. He was a coal miner. He was a good man. I wish he was living today, he could tell you some stories about Alabama and how it was hard down in Alabama, picking cotton and all. In other words, they thought they were moving from a bad place to a better place—and it was a bit better but not much better. He used to sit down and tell us stories about growing up and how hard it was and it was harder when he got here. It got a little bit better but it was still hard. My grandmother, her name was Amanda. She was a good lady but she wasn't from down south.

They lived over in number three. Those were the camps,

the coal camps. Camp number one is where the blacks lived. On Main Street, all them were white. Camp number two and number three, all were black. Number four, they had a place called silk stocking row, and everyone was white. Number five was all black, and in number seven, some were black and some were white. Today, blacks can live on Main Street, but when I was coming up blacks couldn't live on Main Street.

My grandfather used to talk [to] me how they had a hard way to go. They only made about two or three dollars a day. It was hard being a black man, you know. Then John L. Lewis formed a union and that was a good thing, because before they got a union black men didn't have no chance. You worked when they told you to work and you went home when they told you to go home. Once the union got started all that changed.

Now it's all nonunion. The bad thing about that is that the guys that they got working now, when they get too old and can't work no more they have nothing to fall back on. That's really bad. Now in the mines, if

you work and get hurt in the mines they'll fire you and you can't do nothing about it.

This is my hometown [Lynch, Kentucky], I love it. I've tried living in the cities, I didn't like the city, too fast for me. I came on back home.

I was a shuttle-car operator. I would load the coal. They would have a continuous miner that would cut the coal and the shuttle car would pull up behind the miner. The miner would load the coal into the shuttle car, and I would take the coal to the beltline and dump it into a hopper. Then the coal goes up into the beltline. When I first started out I was doing general inside labor, shoveling the beltline. For ninety days you're on probation. When my ninety days was up, first thing I seen was that shuttle car and I liked it. From then on that's all I did was shuttle-car operator. Sometimes it'd be low coal, sometimes it'd be high coal, but when I got down to mine thirty-two it was all high coal. You could just stand up. Some coal was seven or eight feet high. I miss the guys, but I don't miss the work. I really miss the guys.

When US Steel started pulling out they sold the mine to Arch Coal. When Arch came here they had it on the front page of the Harlan Daily Enterprise that they called us all hillbillies. Their main goal was to come here and break up the unions—and they did that. They took it over and they broke up the union. Right now, with the young guys working now, if they even say the word *union*, their job is gone. That's really bad business.

My name is Carl Shoupe. I was born and raised up in Lynch, Kentucky. I was born on November the second, 1946, in the Lynch Hospital. At that time the little town was owned and operated by the United States Steel Corporation. US Steel provided a commissary, that's a place where you buy your groceries; they provided doctors, schools; they provided teachers in schools and theaters for us to go to, recreation—all other type recreational facilities, basketball, football, baseball, all the sports. I grew up a coal miner's son, but the company owned the whole town. I've lived here all my life except for four years I spent in the United States Marine Corps.

The train depot served a very important part in the community. One thing I remember when I was a boy, they would bring the payroll in for the men on this train track. I was in second or third grade, you know. I'd always try to time it to be around when that happened so I could watch them. They'd come out of the car with their big shotguns and bags full of money and stuff. They paid the men in cash, you know. So they come out of the depot and they'd go across that bridge, and that was the main office building right on this side. That

was where they would deliver the cash.

Behind the office was the bathhouse. All the men bathed in there after they came out of the mines. For what it's worth it was segregated back then, you know. Up front here was where the white folks bathed and way back in the back there was [where] the black folks bathed.

Back in those days, US Steel hired the best of men for anything they done. If they needed some kind of special bolt or some kind of special tool made to fit something, they had men in that building that could do it. They had the machinery that could do it. They would make their own electricity and they had a generating plant. The big chimney in Lynch, that was all part of the generating facility. That coal and stuff would go right in there and generate the electricity and ran the town. I guess Lynch was one of the first coal mines to ever have electrical lights inside the mines. Most all the other mines, you know, you had to depend on your cap light or even carbide light I guess, back in those days.

I'm a third-generation coal miner. My grandfather helped start this mine for United States Steel in 1917. My father came to work here probably in the late 1930s and I came to work for United States Steel in 1969, May of 1969. On March 10, 1970, I was working the second shift at what they called number thirty-two mine. I was a roof bolter, so I was bringing up roof supports at the face of the coal. At about nine thirty that night, I was starting a new place and started to put up my first beam when all of a sudden the mountain came in on me. I was covered up for about twenty minutes, I guess. The thing that really saved my life was that there were enough men around to literally pick that rock up off

me. Not only was I crushed real bad, but I was smothered also. Anyway, I only got to work in the mines, literally work in the mines, for about eight months.

After I healed, I got a job with the United Mine Workers Union as an international organizer. My first assignment was the Brookside strike that happened in 1973–74. I was fortunate, or unfortunate, to get involved in that and I got a lot of hands-on experience. Really, if a man needed training to be an organizer for the United Mine Workers that was a good place to start. Anyway, I worked for the United Mine Workers for fifteen years as an organizer.

Then I got on the bottle, got to drinking, and I really became an alcoholic and really went through some trying times. Then on September 26, 2005, my wife talked me into going to church one night and I went to a Trinity Holiness Tabernacle Church in Hiram, Kentucky. That night the Lord Jesus Christ came into my life and since that night I've not taken a drink of liquor or no curse words have came out of my mouth.

The good Lord healed me because when I woke up and I started seeing all this stuff that was going on around me. I saw mountaintop removal, strip mining, radical strip mining, all the dust that was being made and everything. The streams were being filled up with all these toxic materials and stuff. I said, "Man, this ain't right." That's where I started my activism I guess, back.

You know coal came in the twentieth century like a blazing guns and stuff. That was a good thing back then, but here it is the twenty-first century and coal, in my opinion, it's a thing of the past. It was part of my past, and it's part of my present, but I don't see it being part of my future—or my grandchildren's future.

Integrated miners at Wheelright, Kentucky.

Miners' attempts to organize for better wages and working conditions were met with violent resistance, and the Appalachian coalfields were the site of mine wars every decade from the 1890s to the 1930s. During this time period, several unions attempted to gain a foothold in the Appalachians. Most of the conflict centered on the efforts of two unions: the communist National Miners Union and the centrist United Mine Workers of America. After defeats in Harlan and Bell County, Kentucky, in the early 1930s, the National Miners Union dissolved and the United Mine Workers of America became the face of organized labor in the region (Portelli 2012).

The United Mine Workers of America was the first industry-wide union to successfully adopt a policy against racial discrimination (Couto 1993). Union organizers knew that appealing to racial prejudices would play into the hands of the coal operators, who already segregated miners' housing, schools, and churches. Their policy set the stage for other industrial unions such as the United Autoworkers and United Steelworkers to support racial equality as a workplace norm (Couto 1993).

After the labor reforms of the New Deal guaranteed workers' rights to organize, by 1948 the United Mineworkers of America had become the largest union in the country. Almost simultaneously, Congress passed the Taft-Hartley Act of 1947, which greatly restricted the activities of labor unions. This new law, combined with rapid mechanization of the industry during the 1950s, weakened the union. As a consequence, membership dropped to less than twenty thousand working miners by 2014 (Maher 2014).

MECHANIZATION

Technological innovations jump-started by World War II had a dramatic impact on Appalachian coalfield communities. In the post-war years large earth-moving equipment that had been developed for military application overseas was put to domestic use in surface mining (Eller 2008). The widespread adoption of the Joy loader had a particularly profound impact on Appalachia. Although several companies introduced mechanical coal loaders during the

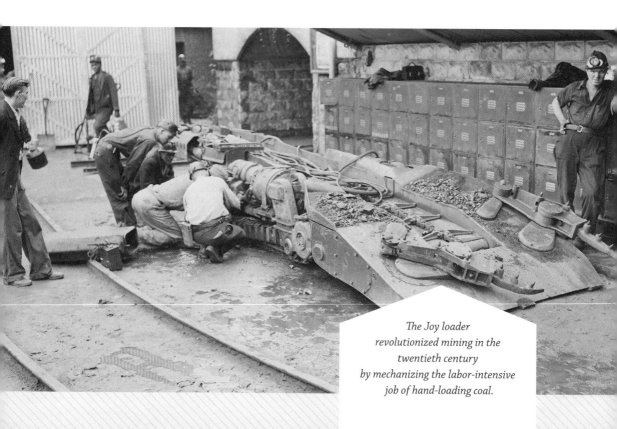

The Joy loader revolutionized mining in the twentieth century by mechanizing the labor-intensive job of hand-loading coal.

MINE DISASTERS

Coal mining is dangerous work and early mines were especially unsafe. Historian Ronald Lewis wrote that during World War I, a coal miner in West Virginia was statistically more likely to be killed than a soldier in the US Army (Lewis 1984). Both south Wales and central Appalachia have the dubious honor of being home to the most deadly mining accidents in the United Kingdom and the United States respectively.

On December 6, 1907, a methane explosion at West Virginia's Monongah mine was amplified by high amounts of coal dust in the mine shaft. The official death count stands at 362 miners, the most lives lost in a single incident in the US. However, mine safety experts estimate that many undoc-umented immigrants from Poland and Italy were killed by the blast, and that the actual death toll may be closer to five hundred (McAteer 2007).

On October 14, 1913, a strikingly similar disaster struck miners working at the Senghenydd Colliery in south Wales. A methane explosion was spread by suspended coal dust throughout the mine, killing 439 miners. Senghenydd stands as the most fatal mining accident in the United Kingdom (Brown 2009).

Above ground, not just miners but entire com-munities have suffered remarkably similar disasters as a result of improper disposal of mine waste. On October 21, 1966, a coal tip (known in the US as a slag heap) owned by the National Coal Board collapsed above the Welsh village of Aberfan. More than 1.4 million cubic feet of debris covered the small village, completely burying a primary school and killing 116 children and 28 adults (McLean 2009).

On February 26, 1972, an eerily similar disaster wiped out entire communities along Buffalo Creek in southern West Virginia. Pittston Coal Company had used mine waste, or slag, to create a series of dams at the head of Buffalo Creek. During a period of heavy rain, these impoundments collapsed, releasing a wall of sludge and coal waste onto the communities downstream. As a result, 125 people lost their lives, 1,121 were injured, and more than 4,000 people were left homeless (Pickering 1975).

Despite the haunting similarities between the Aberfan and Buffalo Creek catastrophes, it is important to note that the government response was markedly different. As a result of the Aberfan disaster, a law was enacted in the United Kingdom regulating the disposal of mine waste, and coalfield residents successfully pressured the National Coal Board to undertake a massive program of cleaning up the waste tips in the valleys of south Wales. After this law was passed, no mine waste disasters have occurred and water quality has improved in these former mining communities (Hansell 2016).

However, in Appalachia, the Pittston Coal Company entered into a confidential settlement with the families of the victims of the Buffalo Creek Flood and no major mine-safety reforms resulted from this disaster (Pickering 1975). Despite decades of activism by coalfield residents to address health and safety issues from mine waste, coal waste impoundments continue to be a common way to dispose of mine waste in the US.

The Aberfan disaster (left). *The Buffalo Creek disaster* (above).

GENDER IN THE COALFIELDS OF WALES AND APPALACHIA

BY PAT BEAVER

"Women in a mining community have it much harder than women in another community. The work is so dangerous that they know they have got to look after the family themselves because their husbands could be killed or so disabled that they die very young. So the woman has got the job of bringing up the children on her own."

— Enfys Davies (Lewis 1976b)

Historic gender roles and relationships in the coalfields of Wales and Appalachia were based on deeply held cultural patterns formed around work in the coal industry. Work in coal is dangerous, difficult, and traditionally male. Although women and children labored underground in the first mines in Wales, they were banned from the mines in the nineteenth century in response to their horrific working conditions.

Welsh trade unions were firmly established by the 1870s. The mines and the unions that supported the miners were male spaces, around which developed particular kinds of male solidarity through shared danger and ideas of masculinity that translated into nonwork arenas, especially the pubs, clubs, and sports (Massey 1994). As a result, opportunities for women to participate in public life were limited.

In the US, the United Mineworkers of America (UMWA) was a largely male organization supporting male labor and male solidarity. Yet the 1970s energy crisis saw an expansion of employment in the US coalfields, including the hiring of women workers, particularly after Equal Opportunity legislation in 1964 and 1972. Women's employment in US coal increased into the 1980s and the Coal Employment Project, established in 1977, provided training and advocacy for women entering the workforce (Women's Bureau [Department of Labor] 1985). In the 1980s, a drop in overall coal employment signaled a dramatic decline in female employment.

In Wales, prior to the miner's strike of 1984–85 women were excluded both from the male world of mining and the male political and social worlds that grew out of the organizations surrounding work in coal. The national miners' strike in the UK in 1984–85 propelled women to the forefront of marches and demonstrations in their attempt to resist the overwhelming forces that would forever alter the livelihood of their families; they moved vast quantities of food, money, medicine, and supplies into the homes of striking miners (Francis 2009). As a result, women developed into community leaders. The vital lines of support provided to the striking miners by the gay and lesbian coalition in London also expanded community ideas about gender and sexual orientation.

As in Wales, Appalachian women still bore primary responsibility for the domestic realm, and found support in extended kin and community social ties. In the Appalachian coalfields, few women found stable work for a living wage, but many held a variety of part-time, temporary, low-wage jobs in the service sector and fast food industries.

Women's involvement has undergirded the workforce. The demands of women's domestic and public responsibilities have typically required flexibility. This means women often are leaders in community development projects, which require both flexibility and the ability to make long-term commitments. With the decline of coal and job loss in the industry, women, men, and families in Appalachia must absorb the social costs of mine closings and economic restructuring (Lewis 1984).

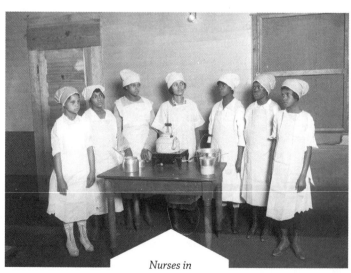

Nurses in Lynch, Kentucky.

1920s, Francis Joy's invention dominated the coal market by the post-war years (*The West Virginia Encyclopedia* 2012). Before World War II, the average central Appalachian mine employed several hundred men, and most of them manually loaded the coal onto belts. A mine with a mechanical loader could produce more coal using only a few dozen men (Eller 2008).

During her interview for the *After Coal* documentary, Appalachian scholar Helen Lewis recalled the era of mechanization:

> I moved to the coalfields in the late 1950s, and people said that the coal industry was dead. Well the coal industry was producing more coal than ever, but they were doing it with machines instead of workers. So while the industry was doing well, coal communities were dying. Between 1955 and 1965, over two million people left Appalachia. Most of them headed north, to industrial jobs in the Midwest.

THE ENERGY CRISIS

The energy crisis of the 1970s breathed new life into the coal industry. OPEC's hold on oil exports increased demand for coal and prices skyrocketed. Organized labor took advantage of the nation's increased dependence on coal to advocate for higher wages and better working conditions. Miners in Harlan County, Kentucky, successfully organized workers at Duke Energy's Brookside mine to join the United Mine Workers of America. Their efforts received national recognition through Barbara Kopple's Academy Award–winning documentary film *Harlan County, USA* (1976). The miners' victory at Brookside became symbolic of the advances gained by organized labor during this decade.

The coal boom of the 1970s brought a brief period of prosperity to the Appalachian coalfields, but it also left scars that are still visible

PAT BEAVER WRITES ABOUT THE EXCHANGE

As director of the Highlander Research and Education Center in Tennessee, Helen Lewis developed an exchange of Welsh and Appalachian coal miners in 1979, funded by the US Department of Labor. That year I was the newly appointed acting director of the new Center for Appalachian Studies at Appalachian State University when I received a call from Helen Lewis. She asked if we could provide a van for the trip and I readily agreed. I highly respected Helen Lewis and her work in Appalachia, and I knew that the university had vans available for educational purposes. As I completed the travel arrangements, I learned that I couldn't simply let Helen Lewis drive away in a university van, but I must be the driver. Thus began a decades-long journey that has taken me to many places and given me new insights on Appalachia, Wales, and the global energy economy.

On a lovely day in May 1979 I set off from Boone, North Carolina, for Washington, D.C., to meet the Welsh miners at Dulles International Airport. We all gathered together in the van and set off, with joyful toasts to the journey ahead and harmony singing. The miners had been chosen for the trip by their local unions, and included Len Jones, a wagon repairer, and Arfon Evans, Ivor England, Ron Rees, Meirion Evans, and Tony Richards, all deep miners. Labor historian Hywel Francis, founder and director of the South Wales Miners' Library at the University of Swansea, coordinated the delegation. After a first-night welcome hosted by West Virginia lawyer and mine-safety activist Davitt McAteer, we headed for Pennsylvania.

Over the course of the next three weeks, the Welsh delegation visited mines, training centers, and health care facilities; made presentations; shared stories; and met with Appalachian miners and their families. There were visits to Uniontown and Centreville, Pennsylvania; Morgantown, Charleston, Beckley, and Gary, West Virginia; and Harlan County, Kentucky, before two days of workshops at the Highlander Center in east Tennessee. We also traveled to Boone, North Carolina, and visited a strip-mining site in southwest Virginia. We concluded the trip with a visit to the headquarters of the United Mineworkers of America in Washington, D.C., where the delegation made a final presentation from the South Wales Area National Union of Mineworkers before returning to Wales.

In 1981 I had the chance to travel to Wales and stayed with miner Len Jones and his wife, Joan, in the Rhondda Valley. During this trip, I saw mining in south Wales firsthand and visited with the other miners who had come to America, as well as with Hywel and Mair Francis. The following year, 1982,

Len and Joan Jones returned to the US to visit us.

Following the rise to power of Margaret Thatcher in Great Britain, and her initiatives to shut down the British coal industry in 1984–85, the south Wales coal communities were shaken to the core. Miners went on strike in massive demonstrations, and thousands of miners eventually lost their jobs. The struggles of that year dramatically changed the south Wales valleys and altered the course of the exchange with Appalachia.

Hywel and Mair Francis visited the Highlander Center and other sites in Appalachia on several occasions during the 1990s. In 1997 they visited the Appalachian State University campus and the adult literacy program at Caldwell Community College. Mair had started the DOVE Workshop in Wales, working with women who were entering the labor force for the first time in the aftermath of the closing of the mines. She was interested to learn from the experiences of others working in community-based adult literacy education.

During the fall of 1997, Helen Lewis, John Gaventa, and I met for a reunion trip to Hywel and Mair Francis's home in Wales. We were joined for an evening with all the miners who had come to Appalachia in 1979, and the next day conducted a workshop on community development issues in the Welsh and Appalachian coalfields at the DOVE Workshop. This was our last reunion with the whole group, our first visit to the DOVE Workshop, and our first formal presentation on Wales and Appalachia in the aftermath of the closing of the mines in Wales and significant job loss in Appalachia.

In 2000 I returned to Wales with the intention of developing a summer field school for Appalachian State University. The following summer (2001), Helen Lewis and I team taught the study abroad course "South Wales in a Post-Coal Society." The course continues today under the leadership of Dr. William Schumann, director of the Center for Appalachian Studies at Appalachian State University.

In the early 2000s, Helen Lewis told me about the collection of videotapes located at Highlander Center that she and John Gaventa and Richard Greatrex had made in Wales during 1975 and 1976. I contacted Susan Williams, librarian at Highlander Center, who located the tapes, and on a visit to Highlander, I collected the Welsh tapes and delivered them to Appalachian State University's Appalachian Collection librarian Fred Hay for archiving. Fred and I were exulting in the possibilities of researching a comparative study of Welsh community life before and after coal, and he was able to find funding within the library resources to have the entire collection digitized. These videotapes provided the impetus for the After Coal project.

The Brookside Mine was the site of labor unrest during the 1970s.

today. The practice of strip mining became much more prevalent as coal operators worked to quickly bring their coal to market. However, cutting into the steep hillsides and narrow valleys of the Appalachian coalfields devastated many mountains, watersheds, and communities. Before the Surface Mining Control and Reclamation Act was passed in 1977, coal companies were free to leave land that had been strip-mined unreclaimed, and large tracts of land remain spoiled to this day.

GLOBALIZATION

During the 1980s, pressure from an increasingly global coal market caused many Appalachian mining companies to reorganize. Companies such as A.T. Massey in West Virginia formed subsidiaries to escape the high labor cost of union contracts. Over the next three decades, the coal industry responded to an increasingly globalized energy market by aggressively lobbying the federal and state governments for deregulation. As a consequence, Appalachian coalfield residents witnessed underground mines closing in favor of mountaintop-removal coal mining, as well as a decline in enforcement of health, safety, and environmental regulations. Deregulation during the 1990s expanded strip mining and the practice of mountaintop-removal coal mining.

By 2000, coal-fired power plants produced more than half of the electricity in the United States. During that year, Kentucky and West Virginia's mines combined to produce more than half of the nation's coal, but that would soon change.

TURNING POINTS

INTRODUCTION

When I started making the *After Coal* documentary, the process unearthed deep-seated memories of watching the six o' clock news when I was in high school and seeing dramatic scenes of thousands of miners clashing with bell-helmeted bobbies during the 1984–85 miners' strike in the UK's coalfields. The most violent footage was replayed during music videos for my favorite punk bands late at night on MTV's *120 Minutes*, which I watched religiously.

Coal-loading facilities at Maerdy, Wales (left), and Totz, Kentucky (right).

While researching the film, I learned that those dramatic strike images from my youth signaled the end of an era for south Wales coal-mining communities. At the same time I was examining the end of coal in Wales, I was also witnessing unprecedented job losses in central Appalachia. This chapter examines the end of coal in both places and provides context for the regeneration efforts discussed in the remainder of the book.

GLOBALIZATION

The big thing that people are saying is that we've got to become competitive on the world market. Well, if we are going to be competitive on the world market with real cheap coal from South Africa, then you've got to make slave labor of your workers here.

—Helen Matthews Lewis, Appalachian sociologist and activist (Donohue 1990)

During the 1970s and 1980s, coal companies increasingly merged with oil companies to take advantage of international energy markets. In both the US and the UK, the coal industry reacted to global competition by cutting production expenses (Donohue 1990). Their first step was to reduce the cost of labor, which presented serious threats to miners' unions. Although the National Union of Mineworkers in Great Britain and the United Mine Workers of America had become powerful political forces in their respective nations, they were not structured to work in an international arena.

WALES

Historian Hywel Francis describes how the British government promoted the idea of a global free-market economy: "Between 1979 and 1985 there were constant attempts almost every year by the right-wing conservative government to reduce the size of the coal industry, and to ultimately break the union and to privatize it all. There were annual battles which culminated in the biggest battle of all in 1984–85—a twelve-month strike" (Francis 2010).

When Margaret Thatcher became prime minister in 1979, she led a massive government effort to privatize industries that had been nationalized after World War II: steel, electric, and most importantly, coal. Coal was the key to the government's privatization strategy because the National Union of Mineworkers (NUM) wielded considerable political power in the trade union movement. If the government was going to take on the trade union movement, they had to fight the miners.

The government fired the opening shot in 1981 when the chairman of the National Coal Board, Sir Derek Ezra, announced a plan to close up to fifty pits, putting about thirty thousand miners out of work (Scargill 1983). Miners in south Wales and Kent immediately went on strike to oppose the plan. When the NUM threatened to expand the strike to other areas, the National Coal Board shelved their plans for the time being and the miners returned to work (Scargill 1983).

During his campaign to become president of the National Union of Mineworkers, Arthur Scargill (1983) described the threat the global energy economy presented to the union: "Much of our market is now being supplied from South Africa, where coal is produced by slave

Miners picket the steel mill in Port Talbot, Wales, to stop delivery of coal.

labour; by short term open cast coal from Australia who have said that their aim is to capture the market and exploit to the full those who utilize their product; and by non-union labor in the U.S.A."

In March 1984, following a landslide election victory by the Tory Party, the National Coal Board announced a new plan to close more than ninety mines throughout the UK (Curtis 2013). A disproportionate number of these mines were in the coalfields of south Wales. In response, the National Union of Mineworkers called a nationwide strike to oppose the plan and to stop pit closures. This would become the longest and most destructive strike in NUM history. After a full year of striking, the union leadership conceded to the coal board's plans. The coal industry quickly privatized, and most mines closed (Curtis 2013).

Historian Ben Curtis describes the results of the historic strike:

> Following the miners' defeat in 1985, the coal industry shrank at a staggering rate. When Labour left office in 1979, there were 235,000 miners in Britain; by early 1992 there were 32,000. South Wales felt this impact as fully as anywhere else. Before the strike, 20,000 miners worked at twenty-eight pits; by 1994,

the government was attempting to shut down Tower, the last
deep mine in the coalfield. These closures were followed rapidly
by the collieries' demolition, resulting in coal mining literally
being wiped off the map (Curtis 2013).

*We lost the strike. The rapid decline of the coal industry, having lost
that strike, created a feeling of helplessness. I personally felt that
it was absolutely soul destroying. There I was, one of the leaders of
the National Union of Mineworkers, and I was completely disarmed
of any ability that I used to have of helping people who were in that
situation. I just couldn't do anything about it.*

—Terry Thomas, former miner and labor leader

The 1984–85 miners' strike provided a dramatic end of coal's dom-
inance in Wales. Simultaneously in the United States, an obscure
strike on the West Virginia–Kentucky border revealed the power of
the global political and economic forces that challenged mining communities in both regions.

APPALACHIA

At the same time in the United States, privately owned coal companies responded to international competition by reorganizing their
corporate structure to escape union contracts. Here is how it would
work: a company with a union contract would sell their assets to a
new company (often with the same shareholders), then this new
company would claim that they were exempt from previous agree-
ments with the union (Couto 1993). These legal maneuverings had
real-world impacts as miners' wages and benefits were cut. In 1984,
the United Mine Workers of America (UMWA) responded by calling

WALES CONGRESS
in support of Mining Communities
WHEN THEY CLOSE A PIT
THEY KILL A COMMUNITY
STOP THEM!
SUPPORT THE MINERS

a selective strike against A.T. Massey Coal Company in order to force the company's newly formed subsidiaries to honor Massey's contract with the union (Couto 1993).

Richard Trumka (then president of the UMWA and currently president of the AFL-CIO) explained:

STRIKE SONGS

HEY, MR. MASSEY
(excerpt of song by Rocky Peck)

Interviewed by Anne Lewis for her film Mine War on Blackberry Creek *about the A.T. Massey strike in 1984, union miner Rocky Peck remembered: "For one day, we had all the trucks stopped. I guess we had about two thousand miners out there, they didn't have no choice but to stop." The scene inspired the miner to write an original song.*

Hey Mr. Massey, we ain't gonna run no coal
We ain't gonna run no coal here today
The union as you know has come a long long
     way
Those old scab truckers that you hired are not
     gonna stand in our way
I believe in our freedom in the good old USA
We're not gonna be pushed around, united
     we're gonna stay
You thought about so many ways to tear the
     union down
But the union's still gonna be here when they're
     putting you in the ground
Hey Mr. Massey, we ain't gonna run no coal
We ain't gonna run no coal here today
You're trying to split our brothers up and have
     your own way
I tell you now it's not gonna work, the union is
     here to stay

BLACKLEG MINER
(excerpt from traditional English folk song)

Strikebreakers, or scabs, were historically referred to as blacklegs in mining communities throughout the UK. This song is a well-known labor anthem.

It's in the evening after dark,
When the blackleg miner creeps to work,
With his moleskin pants and dirty shirt,
There goes the blackleg miner!

Well he takes his tools and down he goes
To hew the coal that lies below,
There's not a woman in this town-row
Will look at the blackleg miner.

They grab his duds and his pick as well,
And they throw them down the pit of hell.
Down ye go, and fare ye well,
You dirty blackleg miner!

So join the union while you may.
Don't wait till your dying day,
For that may not be far away,
You dirty blackleg miner!

A.T. Massey is a multinational corporation. It is a conglomerate of Royal Dutch Shell, the second largest corporation in the world with 76.5 billion dollars in assets and Fluor Corporation, with roughly 6 billion dollars in assets. They both own a coal company, and those coal companies both own A.T. Massey.

Private security guards hired by A.T. Massey Coal Company to break up a strike.

Massey, in turn, has divided itself into hundreds of corporations stacked on top of each other (Lewis 1986).

After a five-year legal battle, a federal judge ordered Massey to
honor its contracts with the union. Although technically a win for
the union, the decision came too late to help the union miners who
had moved on to other jobs as strike benefits were depleted (Couto
1993). Union membership and influence continued to decline, and
by 1999, there was not a single union mine operating in eastern
Kentucky (Pavlovich 1999).

Helen Lewis reflected on these changes:

> People say that we are in a post industrial society, others say
> we are de-industrializing. What we are doing is we are restruc-
> turing older industrial societies. Developed countries are being
> de-industrialized to become leaders in new computer managed
> technology that is industrializing those undeveloped areas.
> Industries are seeking the cheapest products, they are seeking
> the cheapest labor, and they are going wherever they can
> (Donohue 1990).

JUNE 2013: BRYNLLIW COLLIERY, WALES

A breeze stirs the grass of an empty field as Terry Thomas and Eric
Davies walk side by side along a gravel path. They are showing us
the site of the Brynlliw mine, where they started working together
in 1965. As they walk through the vast open space where they once
worked, they try to figure out the location of the mine shaft, the
colliery canteen where they ate, and the pithead baths where they
cleaned up after each shift.

Terry Thomas at his home in Goresinon, Wales.

Terry shakes his head in disbelief. "There were more than 1,000 men working here in 1969; looking at it now you would never guess it was a mine" (Hansell 2016).

Terry had not been back to the site since it closed in 1983. He had left the mine for a job with the National Union of Mineworkers long before it closed, and was working as a union official during the closure. Still the mine closure took a personal toll. Terry told me:

> As a representative of the NUM I had to preside over thousands of redundancies. That was all I was dealing with, as the pits closed, managing the layoffs became my job. All these jobs were well-paid jobs, and they were lost forever. But what could I do? Our people had given all they had to give. People were losing their homes, they were being foreclosed, marriages were breaking up, and suicides were on the rise (Hansell 2016).

The end of coal in the UK was brought about when a globalized economy put pressure on nationalized industries to cut costs.

HYWEL FRANCIS ON WALES AND APPALACHIA: COAL AND AFTER COAL

(a presentation at Appalachian State University in 2010)

During the period between 1979 and 1985 in the south Wales coalfield, we actually moved back into our history, so to speak. We don't refer to them as coal wars, as you do. But between 1979 and 1985 there were constant attempts almost every year by the right-wing conservative government to reduce the size of the coal industry, and to ultimately break the union and to privatize it all. There were annual battles which culminated in the biggest battle of all in 1984–85—a twelve-month strike. And I wrote that history in a book called *History on Our Side* (2009).

Thatcher's Tory government hired a man named Ian MacGregor, who was a Canadian-American right-wing steel manager, and he took over the steel industry in Britain, and he succeeded in privatizing British Steel. And then he took over the coal industry, so it was quite clear that in terms of managerial, quasi-political strategy the conservative government was borrowing from the rich union-busting experience in the United States.

What we witnessed in Wales was a different kind of struggle than in other parts of the UK. And without being romantic or parochial (but I can be very good at both of them, I suppose), what we witnessed in terms of our struggle in Wales was a qualitatively different kind of struggle compared to other coalfields in the rest of Britain. The slogans that were adopted reflected this: "The NUM Fights for Wales" and "You Close a Pit and You Kill a Community." In Wales, we linked the miners' economic struggle to the wider cultural struggle, to the wider political struggle.

This national consciousness added to the growing movement to establish a Welsh National Assembly. The shift in the desire for democratic devolution occurred most dramatically in the areas where the miners had struggled. The Welsh government was eventually established in 1999, and as far as I'm concerned, the demand for that stronger local democracy came out of the miners' strike.

As a result, the conservative Tory government decided that the National Union of Mineworkers was an obstacle to the nation's participation in a global free market and broke the union in order to import cheap coal from overseas. Although coal was still an important energy source for the UK, very little of it was coming from south Wales.

JUNE 2012: PIKEVILLE, KENTUCKY

On a warm June afternoon in the Pikeville city park, Kentucky state representative Rocky Adkins is working the crowd. He shouts into the microphone with the cadence of an old-time preacher: "They have started a war against coal. They have an attack against our industry. They are trying to break the backs of our economy. They have cost our region thousands of jobs. We must stand shoulder to shoulder, and we must fight this fight together."

The crowd cheers, waving signs that read "Got Electricity: Thank a Coal Miner," "Coal Feeds My Family," and "Stop the War on

Rocky Adkins addressing a pro-coal rally in 2012.

Coal." This pro-coal rally is strategically planned to be held next to the Eastern Kentucky Exposition Center in Pikeville, where the Environmental Protection Agency (EPA) is holding a hearing to gather public feedback on proposed regulations to protect head-waters of streams.

Eastern Kentucky has been hemorrhaging coal jobs recently, and many in this crowd believe the industry's claim that the Obama administration's environmental regulations are to blame. An industry-funded group named Coal Mining Our Future has sponsored this rally to support their efforts to kill the proposed stream protection rule.

BACKGROUND

In Appalachia, the turning point for coal came much later and less dramatically than in Wales. Coal production peaked in 2001, but mining jobs had been trickling away for more than a half century (Energy and Environment Cabinet 2016). When the *After Coal* documentary was released in 2016, a combination of declining reserves, government policies, and competition from natural gas created historic lows in central Appalachian coal jobs (Energy and Environment Cabinet 2016).

The power of the United Mine Workers of America quickly declined following the labor disputes of the 1980s. At the same time, strip mining and mountaintop-removal coal mining increased. Despite the environmental destruction caused by blasting mountaintops and filling streambeds, mountaintop removal could produce two-and-one-half times more coal per worker than underground mines (National Mining Association 2007). As a result, between 1990 and 2000 coal employment continued to decline in central Appalachia even as coal production increased.

The weakening of the union also shifted the culture of resistance in coalfield communities. Historically, tensions in the coalfields had focused on union miners' demands for better wages and working conditions from coal companies, who opposed the union. However, as employment declined and more remaining miners worked on nonunion surface mines, the battlefield changed. Few miners dared to publicly stand up to the coal industry. Instead, miners often took the side of coal companies, who aggressively fought environmental groups and government agencies' efforts to protect clean water, clean air, and safety regulations. By 2010, it was not uncommon to see bumper stickers reading "Save a Miner, Shoot a Tree Hugger" on vehicles in the Appalachian coalfields. Meanwhile, the multinational corporations that owned the coal companies moved on, leaving both miners and tree huggers to clean up the industry's mess.

At the dawn of the twenty-first century, coal's future was looking bright in the United States. In the year 2000, coal produced more than half of America's electricity, and utilities had plans to construct one hundred and fifty new coal-fired power plants. However, the economic crash following the September 11, 2001, terrorist attacks and changes in global energy markets meant that few of these plants got off the drawing board. Within a decade, competition from cheap natural gas was undercutting coal on the utility market, and power companies quickly shifted to this low-cost fuel. Between 2010 and 2015, two hundred coal-fired power plants, or 40 percent of the US fleet, were closed as utilities shifted to natural gas and renewable energy (Hitt 2015). By 2016, coal produced less than one third of US electricity (USEIA 2015).

In addition to lower fuel costs, the threat of enforcement of government policies (or new interpretations of existing policies) to reduce emissions also pushed power companies to switch from coal to natural gas. Natural gas produces less carbon and no mercury, sulfur, or other pollutants regulated under the Clean Air Act. After the election of

President Obama in 2008, tensions in the Appalachian coalfields increased, as the new administration instructed the EPA to tighten enforcement of both the Clean Air Act and the Clean Water Act.

And if the combination of global competition, union busting, cheap natural gas, and government enforcement of environmental regulations were not enough to wipe out coal jobs in Appalachia, geology intervened. Today, after a century's worth of coal production in central Appalachia, all of the easily accessible coal seams have been mined. Many of the remaining coal deposits are narrow seams that require specialized equipment. These high extraction costs make much of Appalachian coal unprofitable to mine.

Ken Troske, professor of economics at the University of Kentucky, told the *Lexington* (Kentucky) *Herald-Leader*: "'I put the range of possibilities at somewhere [from] a continuing small decline to a small growth. . . . We are not going to recover the coal jobs we've seen in the state lost in the last decade'" (Desrochers 2016).

CONNECTING THE PAST TO THE PRESENT

In late October 2012, the Appalachian Center at the University of Kentucky in Lexington sponsored the Global Mountain Regions Conference, which provided the opportunity to bring Welsh miner and labor organizer Terry Thomas to Kentucky. This conference brought together scholars and activists from Ecuador, Wales, India, Mexico, Italy, Pakistan, Indonesia, the People's Republic of China, Sri Lanka, Mali, Canada, and the United States to compare experiences, analyses, and practices across mountain regions around the world. After the conference, Terry Thomas and I traveled to Harlan County to meet Carl Shoupe. I recorded their conversation for the *After Coal* documentary.

CARL SHOUPE: Terry, buddy, welcome to eastern Kentucky, I'm just proud to have you aboard here. I'd like to show you some of the landscape and some of the historical places here in a little town called Benham, Kentucky.

TERRY THOMAS: It looks to me as if the history of the coal industry in Kentucky has followed the same lines as the coal industry in Wales. We've both had a hard time and had to fight for every little thing we ever had.

CARL SHOUPE: We're in Harlan County, Kentucky. It's called "Bloody Harlan" because a lot of the union guys lost their lives here. Matter of fact, if you look right down there (points across the street to the Kentucky Coal Mining Museum) that used to be a commissary. That's before my time, but I heard stories from my father and some of the other older fellows who have passed on. They had a union organizing drive going on here at the little town of Benham. They told me the UMWA organizers came in on a convoy, had a bunch of cars and everything. And the company guards had a machine gun set up in the top of that commissary down there. Right down there where the road forks, there was a wide place in the road. When they got that far, the gunfire started. A union organizer named Paul Hodges got shot and he died right there in the street. They said he just bled out right there in the street because nobody could get to him to pick him up.

At one time in Harlan County there was about one hundred and forty-five operational mines. All of them, at one time, were under the flag of the United Mine Workers of America, except one company and that was right here at Benham where we're standing. International Harvester Company, from the day they came in here in 1911 until the day they quit, it was probably about 1962. They never was United Mine Workers but they had their own little company union. Lynch, the next little town up, has always been union. That's where I was raised at. We'd go out on strike and they'd keep on working. When the union went back to work, and we got our raise, the company down here would give them a few cents on the dollar more, to keep the union out that way.

TERRY THOMAS: We are much the same in Wales because during our last strike, which was 1984–85, we were out a whole year on strike. There was a breakaway union in the Nottingham area of Britain who called themselves the Union of Democratic Miners. There was nothing democratic about them, they were just another form of company union. They were siding with the boss each time against the union. Really, that would cause a strike to fail.

Retired Welsh miner and labor leader Terry Thomas (left) *meets retired Kentucky miner Carl Shoupe* (right).

CARL SHOUPE: And you know, these companies they know how to do that. They'll hire scabs. We call strikebreakers "scabs" here in the United States, you know.

TERRY THOMAS: Scabs or blacklegs, yeah.

CARL SHOUPE: I've never heard that, blacklegs? I've never heard that.

TERRY THOMAS: There's this folk song "The Blacklegged Miner" that explains all that.

CARL SHOUPE: These were all coal camps, you know, and so companies like International Harvester owned the town, they owned all the homes. Same way up at Lynch, US Steel owned all the houses. If you worked for them, you paid them so much a room for your electricity, so much for your sewage, or whatever. They had the crews that would come and clean out your outhouses. We had our own theaters, we had our own hospital, we had our own commissary. You went in and mined the coal and they paid you a wage, but that wage just kept circling around, circling around, circling around. You didn't get anything.

TERRY THOMAS: The mines owned the shops as well?

CARL SHOUPE: Oh yeah! Exactly.

TERRY THOMAS: The company stores.

CARL SHOUPE: So have you been involved in this activist stuff for many years?

TERRY THOMAS: I started in 1960 when I went underground to work in Wales. Almost from the start of my time there I became involved in the miners' union. I was involved in the local section first of all, and then I became the branch secretary of the local mine, and then I was elected to the executive of the National Union of Mine Workers in Wales. From then, I was elected to be the vice president of the union in Wales. So I've been very involved in the union, and all the time I was involved in the union I was involved politically with the British Labour Party.

CARL SHOUPE: Yeah, the Labour Party, I think they're out over there.

TERRY THOMAS: Yes, they are out at the moment and we have got a right-wing Tory government in place.

CARL SHOUPE: Right. Margaret Thatcher, man. Wasn't she . . . ?

TERRY THOMAS: She was the one who was the prime minister during the miner strike of 1984–85.

CARL SHOUPE: Right.

TERRY THOMAS: But if you go back a little bit further in history to 1974, the miners were on strike and Edward Heath was the prime minister. The miners were on strike, so Heath called an election on the slogan "Who Governs Britain?", and the people of Britain voted him out, he lost the election. The Labour government was elected in 1974 during the miners' strike and the miners did win that strike and they gained a great deal. But it was only temporary, because when the Tories got back in, when Thatcher got back in 1979, she was preparing all the time to really destroy the miners. She was having her revenge for what happened to the Conservative Party in 1974, and the miners would come out on strike because of what Thatcher was doing to the coal industry. We were on strike for a whole twelve months.

CARL SHOUPE: Yeah, I remember.

TERRY THOMAS: And we were fighting on the picket lines, and when you tell me about people getting shot on the picket line, I must say that I'm pleased that the British police are not armed. I'm pretty sure that the things would have been much worse had the British police been armed. What they did do was to run miners down with their horses. They beat up miners on the picket lines, but they didn't have guns to shoot us with.

CARL SHOUPE: You know, nowadays there's no guns and stuff but they figured out ways to beat you with a pencil. That's the way I explain it. It's still a battle, Terry, you know? You'd think when you get older, and you retire and stuff, but there's no rest.

TERRY THOMAS: The latest crisis, the latest economic crisis, is a financial crisis caused by the bankers, caused by greed of the bankers. But who's paying the price for that again? It is people in communities such as this paying the price for that, for the depression that was caused by the banks.

CARL SHOUPE: So there's not a whole lot of mining, per se, that's going on in Wales?

TERRY THOMAS: I tell you this, after the 1984–85 miners' strike in Britain, we lost eighty-five thousand mining jobs. At one time, I'm going back in history, back in 1919, there were almost a quarter of a million miners working in south Wales alone. As we're talking now, we've got about six hundred. That's in total, maybe about six hundred. At one time there was a quarter of a million people working in the mining industry.

CARL SHOUPE: Does the state still own those smaller mines?

TERRY THOMAS: No, no. Thatcher's government privatized them. They were nationalized in 1947 and they remained nationalized until the 1990s, but it was the Thatcher government, the conservative government, that brought about the privatization of it. The rail industry, the gas industry, the electricity industry, the coal industry, they were all publicly owned. The principal of nationalization is to produce something for the benefit of all the community, not for the benefit of a few shareholders. That has all been destroyed by the Tories, by Thatcher and the conservative government in Britain.

CARL SHOUPE: She stayed over there a long time, didn't she?

TERRY THOMAS: From 1979 until 1990. I think that people have got to look back at their history, because in mining communities the people in Wales and I'm sure the people in the US of A know that their life was improved by the organization of the trade union. Their life was improved by the union. Now that that life and that industry has been taken, they have to go back and use the methods that were used by the trade unions to improve people's lives in the community. Self-education, building of their welfare halls, cultural events, all those things did bring about an improvement in life, but they were brought about by the union. Now, what we have to find is an alternative to that. We have to find an alternative way to bring about improvement within our communities, because if we are waiting for the politicians to do it for us, then I'm afraid that help is not going to come.

Wind turbines loom over the coal washery in Onllwyn, Wales.

CHAPTER 4

EXPLORING REGENERATION

TRAVELING TO WALES

Once I had gained an understanding of Welsh coalfield history, I focused on the logistics of traveling to Wales to record post-coal stories. As I worked with project advisors Hywel Francis and Mair Francis on our itinerary, I wondered what I would find when I got there. The simple fact that Welsh communities had survived gave me hope. I naively imagined that I might be able to discover the magic formula that allowed Welsh mining communities to rise from the ashes of industry, bottle it up, and bring it back to Appalachia.

Our crew of co-producer Patricia Beaver, cinematographer Suzanne Clouzeau, and myself landed at London's Heathrow Airport in May 2012. We rented a small van and headed west on the M4 motorway. Once in Wales, the landscape quickly changed; craggy hills rose steeply on the right of the motorway and Swansea Bay sparkled in sunlight on our left (we would soon learn that the dazzling sunlight that welcomed us to Wales was, in fact, a rare sight). We passed under the smokestacks of the steel mill in Port Talbot, then left the motorway at Neath to enter into endless orbit around the set of roundabouts that pointed us up the Dulais Valley. Ten thousand miners once worked in that valley, but the only sign we saw of coal's legacy was the Cefn Coed Colliery Museum, which was closed for renovations. Jet-lagged, we arrived at our destination in time for a meal and a walk, then much-needed sleep.

The next morning, we interviewed project advisor Mair Francis about her experience founding the DOVE Workshop. "The DOVE," as locals call it, is an organization led by women that provides learning opportunities and space for social enterprise. After the interview, Mair traveled with us up the valley and introduced us to the women who currently run the DOVE Workshop in the village of Banwen. Over lunch with DOVE's co-coordinators, Lesley Smith and Julie Bibby, we arranged for an on-camera tour and interview the next day. The well-ordered efficiency of DOVE's building and the warm welcome confirmed my view that Wales was recovering from the collapse of the coal industry, but a few clues reminded us of coal's legacy in the valleys. A coal train ran up the valley twice a day carrying imported coal to a nearby washery (known in Appalachia as a tipple), where it was mixed with coal from a local opencast mine and shipped to the Aberthaw Power Station on Limpert Bay. Although coal was no longer king in Wales, it was far from dead.

FIRST INTERVIEWS

When interviewing people in the valleys of south Wales, I was keenly aware of my place as an outsider. Although I was not raised in the Appalachian coalfields, I had worked there for decades and had developed techniques to navigate the complexities of recording interviews from this often misunderstood region. Now in Wales, I struggled to learn a new set of complexities. Mair Francis was a guiding force who kept our crew from making too many gaffes. Still, being a stranger with a camera in a place often misrepresented in the media is a challenge that requires constant vigilance.

The difference in language added to this challenge. Just as visitors to Appalachia take time to tune their ears to vowel sounds and speech patterns specific to the place, visitors to Wales need time to tune in to the local sounds. Over time I came to enjoy the lyricism of a Welsh accent, but at first, it seemed very foreign to me. And it was not just the accents that were foreign. Concepts central to community life in south Wales were also new to me. Ideas about mutual responsibility and how people work together were markedly different than my experience in Appalachia. The following topics highlight a few of the concepts that kept coming up in interviews we conducted for *After Coal*: community regeneration, social enterprise, and community funds.

COMMUNITY REGENERATION

As I interviewed community development workers in Wales, many people threw around the term *regeneration* as if I would know exactly what it meant. It took me a while to understand *community regeneration*, so I'll explain it first, then get back to the story of our first trip. During our second week in Wales I asked Victoria Winckler, director of the Bevan Foundation, a Welsh think tank of sorts, to define it for me. Here is what she said:

> *Regeneration* is one of those words that means a lot of different things depending on who you talk to. I think a spectrum of different issues and different approaches belong under that umbrella. In one extreme, regeneration is what someone described to me recently as "ethical property development." So you would take a rundown area, very often an urban area, and with a mixture of incentives you attract private investment and lo and behold, property values shoot up. With this approach, the property might look nice, but it's a big question mark whether people have actually benefited. Then, there's kind of a middle ground, where public sector might be trying

Victoria Winckler, executive director of the Bevan Foundation, speaks to the After Coal *crew at the newly renovated Ebbw Vale steelworks.*

to attract businesses to relocate to an area or to develop and grow an area. Again, that's the usual mixture of property and incentives. This approach is actually really important if you've got an area which has had such massive decline to show that these areas do have potential, that they are a decent place to do businesses. Finally, there is the third element of regeneration, which is quite commonly used in Wales, and that's around community and social regeneration.

There's an understanding that regeneration (or any sort of significant change) only happens when people change.

BEVAN FOUNDATION

The Bevan Foundation describes itself as "an independent, non-political think tank, which develops ideas to make Wales fair, prosperous, and sustainable" (Bevan Foundation 2015). The organization takes its name from Aneurin ("Nye") Bevan, who is a Welsh hero. He was born in Tredegar in 1897 and worked in coal mines from a very young age before becoming a labor organizer and politician. He became a leader in the post-war Labour government and helped establish the National Health Service, which provides free health care for all citizens.

In Wales, people have had a lot of difficulty adjusting to the process of deindustrialization and a lot of social relationships have been really damaged by that process, within families, but also between neighbors in communities. There's a belief, a quite widely held belief, that the way you get social and economic change is if communities themselves campaign for it, so it has to be a real bottom-up movement. And, to its credit, the Welsh government recognized that and historically put quite a lot of money into community regeneration projects across Wales in a program called Communities First. That program involved putting particular interest in south Wales because of the regeneration issues.

Victoria Winckler continues on p. 92

Looking over the village of Banwen toward the Brecon Beacons in south Wales.

MAIR FRANCIS ON THE HISTORY OF THE DOVE WORKSHOP

I was the founding member of DOVE, which stands for the Dulais Opportunity for Voluntary Enterprise. Our story is the story of a group of women who, during the miner's strike in 1984–85, embarked on a journey that changed our lives forever, and made a lasting impact on other women and men in our community.

When the strike ended in 1985, most of the miners' wives were exhausted from worry and despair. But a small group of us continued to meet. The support group we formed during the strike provided us with experiences that led to the creation of DOVE.

The women in our support group became active on the political stage during the miners' strike. We spoke alongside union leaders and politicians (who were usually male), marched in demonstrations, and picketed. We met other political groups of women who were involved in the peace movement, such as the campaign for nuclear disarmament, and during this time we had the opportunity to meet feminist writers as well as leaders in the gay and lesbian movement.

These political experiences allowed us women to learn how to work collectively with others. We had a democratically structured committee—which was very important because we needed to have women be in control of our own thoughts and ideas. Through the support group, women were given the opportunity to become politically active.

The primary objective of the DOVE Workshop was to improve learning opportunities for women in the valleys. Women in our area were given secondary roles, they were invisible. All of the organizations within the valley were controlled by men, men were the voices. So, women had a backstage role, and there was no real encouragement for the women to come forward and take control. And so we knew that when we set up DOVE, it had to be run and organized by women for women. And there were criticisms—you know, what about the opportunities for men? Well, men have always had opportunities, let's give the women opportunities first.

We had to overcome many barriers to achieve our vision. We knew from the start that we had to provide free childcare facilities. We couldn't expect the women to come to an educational training program if we didn't provide childcare. It had to be a flexible learning environment, so everything had to be part-time and within school terms, and any holiday period there were no courses running. And of course, we had to provide transport. We were in a rural area. The bus service was nonexistent. Forty percent of the people had cars. So we worked out a way to provide transport. Our first minibus was the van that was donated by the gays and lesbians from London to the miners during the strike. So when the strike ended in 1985, we took over using that van in order to get around to meetings.[2]

Next, we had to find a building. At the beginning we used a temporary building that had been a pool hall. The building that we eventually were able to use was the opencast building that was owned by the National Coal Board. This

Upper left: *Mair Francis, one of the founders of DOVE Workshop.*
Above: *Women march in support of the miners during the 1984–85 strike.*

building was in Banwen, which is a small village of one street at the top end of the Dulais Valley, and one of the most deprived wards in the area. There was a large room for the crèche (day care). That was the one exciting thing about it. There were rooms for seminars and meetings.

Once we had a building, we had to identify tutors. We didn't want a top-down approach. It was a grassroots approach. We wanted our women to become the tutors. So we found tutors that were sensitive to the needs of women. Then we needed funding. At the beginning we were able to get some small grants for voluntary groups. We set up our committee and applied for various bits of money, but we failed many times before we were successful.

The building also had extra offices that became accommodations for the extra-mural learning department of Swansea University (which was directed by Hywel Francis). This department became known as the Department for Adults and Continuing Education, and this was the first time that a city-based university was locating

itself in the valleys on a permanent basis.

Eventually, we started offering flexible, part-time courses as a way of inviting women in and making them feel confident and giving them the opportunity to do video production and learn computer skills. Through this work, we were able to build links with the local college in the town of Neath and with the Workers Educational Association to develop a curriculum so that women could build skills and opportunities as they advanced from one course to another.

During this period, we never lost our aim and our passion to set up a cooperative. We knew that we couldn't rely entirely on public funds. We wanted to develop ways to earn money, which would help us fund the rest of the activities in DOVE. At that time, the main planks of the cooperative were video production, desktop publishing, the café, and the nursery. We figured out how each of these activities could earn income for DOVE.

And, of course, all of this development with the cooperative became the set for the social enterprise initiatives that

are so successful in DOVE today. Now, the social enterprise arm covers the DOVE day nursery, the community garden, and the café. And, at this moment, the community garden is developing into a trading arm. They sell the produce from the center and the local marketplace also sells the produce, not just fruit and vegetables but preserves—jams and stuff like that. And the DOVE day nursery now provides an after-school club for primary school children. They've got a pick up and drop off at schools service. And the café has just expanded to a fifty-seat restaurant with a state-of-the-art kitchen for catering.

Eventually DOVE was seen as an example of best practices for women-specific training. We developed techniques for women to learn alongside women, and we identified women that were coming through on courses to become peer tutors. There would always be one woman in a course that was a good communicator, bright and feisty and intelligent. And we would ask that woman: "Would you like to try being a tutor? We can send you on a mentoring course to university, we'll give you all the help and support that you want."

The two women that run the center at the moment, Lesley Smith and Julie Bibby, are examples of two women that came through as volunteers and went on courses, got more and more qualifications. Last year, they both qualified for their master's degrees in lifelong learning. So, those are real success stories. But there are many success stories like that.

The DOVE Workshop in Banwen, south Wales.

Victoria Winckler, continued from p. 89

We've seen a massive, massive loss of jobs in south Wales over the years. For example, Merthyr Tydfil, which is not far from here, lost half of its jobs in about thirty years. Some communities, smaller communities, have lost all their jobs. If you had almost all of the community working in a coal mine, when the coal mine closes, there's nothing. Our take on that is that it's a matter of social justice. We don't think that people should necessarily get up and move or that the fact that they're unemployed is because they're lazy, or unskilled, or stupid, or any of those things that are said about people who have lost their jobs. Our view is that employers have taken those jobs away from Wales in pursuit of bigger profits elsewhere, and that caring for people and making sure people have a decent quality of life is a collective responsibility and one that is a hallmark of a decent and civilized society.

I think the government has a responsibility to help communities to adjust to the aftermath, but I think that responsibility is about facilitating. I don't think it's about directing or forcing or imposing. I think it is the role of government to support and to enable people to come to terms with the changes themselves, and to create their own alternative futures.

I would say there are lessons to learn and I think one of the biggest ones is that it takes a long time. There are no quick wins. You can possibly have some quick wins around the environment or improving a building or something, but coping with the consequences of change takes time. I think it is inevitably painful. I think it's just a question of just how much pain there is.

—Victoria Winckler, director of the Bevan Foundation

REGENERATION AND DOVE

When we arrived at the DOVE Workshop the next day, we quickly saw a living example of a grassroots approach to community regeneration. DOVE is located in the village of Banwen (population four hundred) at the head of the Dulais Valley. To get there, you wind through several former mining towns before emerging on a high plateau with a view of the Black Mountains and the Brecon Beacons National Park. The wide-open sky and relatively high elevation make Banwen feel more like Wyoming than West Virginia to my American eyes. DOVE is situated at the end of the road, but clear signage in Welsh and English leads you there. The workshop is housed in the office of the old opencast mine that shut down two decades ago. Today, the project is run by a group of women with a vision for a future beyond coal.

Julie Bibby welcomes the After Coal *crew to DOVE Workshop.*

We always felt our dreams could become reality, and those dreams are there now, they are real.

—Mair Francis in the *After Coal* documentary

DOVE co-coordinator Julie Bibby met us outside. We had arranged for her to give us an on-camera tour of the building. That morning, we had decided to use Suzanne's Steadicam to film the tour. I agreed to the plan without understanding the time required to set up a Steadicam rig. The rig requires a delicate balance in order to achieve the smooth floating motion that the Steadicam is famous for. The fact that Suzanne had not used our camera with her rig meant it took extra time to set the counterweights and achieve perfect balance. Fortunately, Julie was clearly used to talking to media and very patient and polite. The fact that we were traveling with Mair and that they knew Pat helped tremendously. Finally, we were ready to record the welcome scene.

Camera and sound rolling, we walked up to the door of the DOVE Workshop, a low concrete building with hills of pine forest stretching up behind it. The camera focused on the symbol of the DOVE Workshop, then right on cue, Julie Bibby opened the door and welcomed us inside. As I stepped through the door, my microphone got caught in the doorframe and I quickly moved it away from the door, causing the boom pole to crash into Suzanne's head. Fortunately, we all laughed. After the comedy of errors settled, we finally recorded the scene. After all this effort, it turns out we were unable to use this scene in the film. However, Julie's story is a powerful one, and I am glad that we are able to include it in this book.

JULIE BIBBY: Good morning and welcome to the DOVE Workshop at Banwen. I'm Julie Bibby and I've been working at DOVE since 1987. I'd like to welcome you in today.

TOM HANSELL: Start from the top.

*Lesley Smith
(see p. 97) at her
desk in the DOVE
Workshop.*

JULIE BIBBY: The DOVE Workshop was set up during the miners' strike of 1984–85 by a group of women who supported their husbands [while they were striking] and thought that they needed to have a place to train so that they could earn some money. They took over this building in 1986.

The building originally belonged to the National Coal Board but was gifted to the Community Council, and the council just took a peppercorn rent from DOVE when they occupied the building. The main aim of the organization is to provide education and training to adults from sixteen years up, so that they can take courses in their own community. We arrange courses from "return to learn" to art classes and craft classes all the way through to a part-time degree in humanities, which is offered by Swansea University.

We offer rooms for rental for private companies, organizations, community groups, and we have a wide variety of activities that go on throughout the week and around the year. We have twenty-four members of staff. We have a day nursery on-site, which is registered for twenty children from zero to five. We employ seven nursery workers, a nursery manager, and the nursery is self-sustaining at the moment.

Julie holds the door open and offers a polite invitation: "Yes, if you'd like to come in . . ." Once inside, she shows us a large black-and-white photograph on the wall of the café:

> This picture was taken before renovations, when the opencast mine was just working alongside us. We had to come in the morning and dust everything down before we could start work. Everything was covered in a fine dust of coal. So you can see the difference in the building today; how nice and green it is outside and the different look of the building . . . it doesn't look industrial anymore . . . it's more pleasing to the eye.

Still talking, Julie starts walking toward a door in the back of the café:

> The café was designed so it leads us right through into our garden. This is our learning garden; it was created to encourage people taking horticulture classes to come and work in the garden. The nursery has got their own gardens, and they've got their own raised beds where they grow their own vegetables as well. So the learning theme starts when they come to the nursery and continues all the way through the courses we offer to adults.

After the walking tour with Julie, we arranged for a sit-down interview with her co-coordinator, Lesley Smith. Lesley shared how her idea of regeneration had evolved from thinking of it as simply a buzzword to understanding the concept as part of a strategy for long-term community development. Yet, she cautioned us to be careful how we use the word *regeneration*, explaining that it can be

PERSONAL HISTORY: LESLEY SMITH

I grew up in Crynant, one of the smaller communities in the Dulais Valley. My father was a coal miner and Mother was adamant he couldn't stay in the coal mine forever because of the health issues. So he left and we moved around a bit. My husband is from Swansea and we lived in Swansea at the time. I'd done a number of jobs. I had nursed, I worked in shops, never settled, never found the right thing, you know. Then I came back to the valley because it was safer for my children. I wanted to be back where I belonged, I suppose.

It was a bit of a culture shock for my husband. People say to me all the time: What's in the Dulais Valley? Well, I don't know what it is, but there's a sense of community here that's really important to me. I don't want to be in a town where we're anonymous and people don't know my children. . . . It's a very warm community. It has got its problems, you know. Sometimes you can feel quite claustrophobic, because everybody knows (or thinks they know) everything about you. But there are lots of things that compensate for that, really.

I came back to this valley in—I've got to work this out now—in 1986. Not for one minute thinking I'm going to get work in my valley. I had two young children, had no childcare, I was looking after my grandparents and my husband came from Swansea with me. And for the first couple of years I helped out in the school, I did what mothers do, got involved in my children's education and just did a bit around the community. And then one day I was in the local library, was volunteering in the local library, and somebody said you've got to come to the DOVE Workshop with me. They are doing great work up there, they have got great courses on, come with me. And I met my colleague Julie for the first time. She was running a taster program, a sample of sorts . . . So I came to that and from that point on I haven't been home!

I really engaged in the organization straightaway. I felt at home, you know. I felt there was a purpose to the organization. I met Mair, volunteered for the employment scheme, and here I am. I started off as somebody who had never typed anything, who didn't have any skills or learning in any shape or form. But, I had an avid belief in what learning could do for somebody—and loved it myself. I've quietly gone through the ranks. I did a degree here, earned the BA in humanities. Then, two or three years ago, I finished my master's in lifelong learning.

PERSONAL HISTORY: JULIE BIBBY

As I was a girl being brought up in this area, in this village, we had thirty-five shops and now we have two. We used to have two petrol stations and now we don't have any at all. So as a child, I had a lot of places to shop then, but then I think at that time everybody was mining and working in the mines, so everyone was the same . . . now is different because the mining is gone. There's no employment here and people either drive out of the valley to work or are unemployed. This is where we are here to help, to move people forward or to help them with the changes in the benefit systems.

When I first came back home to Wales, I thought: "Oh, how will I cope?" But, I found a job at my doorstep, which is good. The children went to local school, we had the day nursery here so I wasn't tied down. Although my parents were good for helping me with the children, I didn't have to rely on them solely; I had my other family, the DOVE Workshop.

I think I came back at the right time because the DOVE Workshop had only been set up. I thought, "I'll take the girls now to the mother and toddler group," and that was my first step into the organization. Then I met Mair and she said: "Are you Julie Jackson?" (which was my maiden name), and I said: "Yes," and she said: "You can type and do payroll," and I said: "Yes," and she said: "Oh, lovely." And I managed to get on a government scheme then just to work twenty hours a week, just doing a bit of clerical work, and I used to drive students in and out with a cranky old minibus we used to have. I never went home really, I just stayed and developed and got more involved with the work.

I worked my way through the organization and became a tutor for the university and the college, and I am now one of the joint coordinators who run the organization. So I'm coming up to my twenty-fifth year at DOVE. We all live local and I think people recognize us and the work we do up here, and I think we've gone from strength to strength through the hard times, which we are going through now with changes in government, changes in funding. So it's always a challenge, but I think we always step up to the mark and continue to carry on our good work.

used to shore up the status quo instead of addressing inequities left behind by extractive industries such as coal:

> When DOVE was established, it was very much in the style of a lifelong learning center and we are still called that by a lot of people. But we came to recognize that the work we were doing (it probably wasn't even intentional) has been regenerating communities and regenerating economies.
>
> It took a long time for us, I think, to recognize that regeneration actually was happening because we were focusing primarily on individuals at the time and not looking at the wider consequences. So I wouldn't say it was by accident, but it wasn't our primary aim. But regeneration has become our primary aim because we see the benefits that we bring.
>
> Later, *regeneration* became a buzzword in government. If you wanted them to fund your project, you would say it will regenerate this area. And without even really knowing it we were starting to do that. When we work with people, when we develop projects and employ people, we add skills to our community and that has an impact on the economy. The people we work with get jobs they get didn't get before. They are better paid and the children see them as role models so the regeneration is going on.
>
> Physical regeneration and environmental regeneration may be helpful in the short term. But, we've also got to remind people that, while mining wasn't a good industry for some, it was a very well-paid industry in the end. Miners earned high salaries with good pensions, and what we do is not bringing that to people. You know, the employment that we offer just doesn't pay as well as mining. It doesn't give as much benefit, so we've got to be very careful about how we term regeneration. It's okay for a place to look pretty, but there's got to be other economic benefits as well.

SOCIAL ENTERPRISE

Through the women of DOVE, I also learned about the concept of social enterprise. After returning home, I decided to look up the definition of the term. The Social Enterprise Alliance (2017) defines a social enterprise as "an organization or initiative that marries the social mission of a non-profit or government program with the market-driven approach of a business."

For DOVE this means that their nonprofit grant-supported classes and workshops bring local people to the workshop. In turn, these people support for-profit enterprises like the café and nursery (the women of DOVE call the nursery "the crèche"). The crèche is a vital

A gate to the park in Seven Sisters where the colliery was located.

element of DOVE, as it allows parents, especially single mothers, to attend classes without worrying about childcare.

Lesley Smith explained:

> We eventually formed a social enterprise called the DOVE Workshop, Limited. Because we're a charity, we have limited abilities to earn income. So the social enterprise arm of DOVE allowed us to trade. We started with childcare and then we looked at a small catering business which went by the by. And yet with the café, here we are again!
>
> Today the café is making some money, but more importantly, it is supporting people to work in the community and get skills in the community. It has become a very important

After Coal *director Tom Hansell and cinematographer Suzanne Clouzeau interview Geraint Lewis at the Maes Y Fron farm in Abercraf, Wales.*

meeting place for people, you know. In this community, unless you go into the club or pub, there is nowhere to sit to meet with your neighbors and have a chat. You can't stand in the street because it rains too much!

The only hope I can think of is that we need to look at indigenous small businesses. We need people to be more entrepreneurial. Because when you get people to run a little business in the community, they're not going to get up and walk away when the going gets tough, you know. They are grounded here. So, it's not always about making a lot of money for one person, it's about making a living, which is a completely different thing. And I think that's what we're really aiming for. We're looking at any opportunity to create a little bit of business that can sustain one, maybe two people.

CALL OF THE WILD

Just down the valley from Banwen, where the DOVE Workshop is located, lies the village of Seven Sisters. The village was built around a coal mine that operated until 1963. Today the only sign of mining is the winding wheel in the park that was once the site of the colliery. In Seven Sisters we visited Geraint Lewis at his business, Call of the Wild, who told us about his company's approach to social enterprise.

I first met Geraint outside of the Onllwyn Miners' Welfare Hall, as he was heading in to choir practice. He has sung in the choir since he was eighteen and his father, nicknamed Big Al, has sung in the choir most of his life. Standing outside the entrance to the miners' hall, Geraint explained that the choir was part of the coal industry's legacy:

> I joined the choir twenty-odd years ago. I was probably the only man who wasn't in the coal industry at the time. Likewise, when

I played my first game for the rugby team in the village, the first year I was the only non-miner playing for that team. So, mining was deeply ingrained, and they worked hard together. They also played hard; they had fun, and obviously singing is central to Wales anyway. You know, everybody likes to sing in Wales.

After choir practice, we made arrangements to meet for a tour of Seven Sisters and Maes Y Fron, a farm outside of town that Call of the Wild had recently renovated into a training center. At the farm Geraint told us how his experience growing up in a mining town ensured that elements of social enterprise were an essential part of his business plans:

> I was born and bred in Seven Sisters. I eventually studied astro-physics at the University College in London. I traveled to the States for a bit of backpacking, and then backpacked for about a year down in the Far East and Australia and New Zealand. My main reason for doing that was that I was wondering if there was anywhere better in the world to live than Wales. I was so convinced that Wales was the place for me, but I wanted to just double-check that there was nowhere else that suited me better. And I enjoyed my trip, it was a fantastic trip, but really, when I came back, there was only one place that I wanted to live, and that was south Wales.
>
> Coal is probably the biggest reason behind the develop-ment of a lot of villages and communities in the south Wales coalfield. In my village, for instance, I would say two genera-tions before me, probably everybody was either working in the coal industry or in some related job, heavy industry job. And then my father's generation, again, the vast majority, I would say were miners. For my generation, the pits started closing as we were in high school, and those jobs weren't there, you know, where they just open the gates and you go from school straight into a job, that sort of stopped. A few of my friends were in the

mining industry, but quite a few of us went on to further our education at the university.

So after doing my degree, I came back, found work—which is the usual way for my generation. The ones that did come back generally couldn't find any work in the village of Seven Sisters, so you had to go to the nearby towns of Swansea or Neath Port Talbot. I found a job as a technical support for a large Japanese electronics firm and filtered through the ranks there until I was on the board of directors as a nonvoting member. In a nutshell, I was told by the Japanese managing director that my next move was at forty years of age, and I was twenty-eight at the time, and I decided to do something that I had really wanted to do. That was basically the starting point of Call of the Wild.

We wanted to start this training development company, using the outdoors as our classroom, which is the key. But there was nobody doing it in the south Wales coalfield. We had a lot of raised eyebrows from a lot of business consultants and bank managers, and saying, "Really? Do you really want to start something like this in Seven Sisters?" Yet we persevered, and for a couple of reasons I think we succeeded very quickly.

First, our overheads were a lot less, because house prices inside the coalfield were much less than those outside the coalfield. There was a bit of business help, which also helped. But more importantly, our staff were key to our success. We pretty much knew every single member of staff and headhunted each one individually. We knew them socially, but I think the other thing that shines through with south Walean people in particular: they're very, very friendly people. In any kind of service industry, or any kind of training where there's one-to-one interaction with clients, that friendliness just shines through and people just want to keep coming back. They build up strong relationships and they want to keep coming, and I think we can't underestimate the massive effect our staff have had on the growth of the business.

We do the full range of activities . . . we do everything from climbing, caving, canyoning, kayaking, mountain biking. We do pretty much everything. But it's not central to the company and what the company actually does. Those activities are seasonal.

Of course more people want to go into the outdoors in the drier, warmer periods of the year, but we effectively developed our company to move people along, make people more aware of themselves, their own natural behaviors. And if people can move a little, and understand themselves better, well that's really our job, then. We apply that to whether it's a board of directors of a large utility firm or ten-year-old children. It's the same sort of process, it's putting them a little out of their comfort zone and seeing if there are other things out there.

We think the outdoors is the best classroom there is. If we delivered some of our management programs to our clients in a stuffy hotel somewhere off a motorway, the result would be nothing like doing exactly the same thing sitting under this tree here. It just opens people's minds, they think differently, they think more freely. We've done studies that prove exactly that—the learning is far more deeply embedded when it's done in the outdoors.

So, for instance, with the school programs, we like to use as many activities as possible, because a lot of the children that age won't have tried any of those activities. So the schools' program is very heavily activity oriented. What you'll find with the management groups, quite often they won't do an activity, other than maybe a hill walk or something like that, but that's about it. Unless they want to be challenged and pushed further.

Geraint explained that Call of the Wild's work with local schools is an example of the company's approach to social enterprise. Basically, the services they provide to expensive boarding schools help support programs for local students:

> Well, most of our paying schools come from across the border, in England. But a lot of our services are to the local primary schools. We try and do free days for all the local schools. There are about six local primary schools, so they'll get a free day once a year for their oldest pupils. And we'll take them caving, or something quite high-adrenaline. And again, it's one of those things that supports our community spirit and shows that business can be successful here.
>
> I think we're bucking a trend that started a long time ago, which is now starting to reverse, I think. I can walk to work, and pretty much all our employees can walk to work. That's something that hasn't happened in Seven Sisters for a long time, because once the mine closed, all the jobs were fifteen or twenty miles away. And obviously, the fact that we employ thirty people, that's quite a big employer for the valley; when you've got a village size of about two thousand people, that's quite a chunk of people employed.

Talking with Geraint Lewis also helped me understand the importance of natural assets to building a local economy. Clean air and clean water are key to getting any kind of business to put down roots, and the reclamation of mine sites in Wales provided important assets for regeneration.

AFAN VALLEY AND COMMUNITY FUNDS

From Seven Sisters, we traveled down the Dulais Valley, then down the Neath Valley to Swansea Bay, where we hopped on the motorway briefly before exiting at Port Talbot and turning up the Afan Valley toward the village of Glyncorrwg.

The steep ridges and tight valleys of the Afan Valley felt more like home to me. I had to keep reminding myself to drive on the left-hand side of the winding two-lane road. Many of the hillsides bore scars of recent clear-cutting, and this obvious evidence of extraction was another connection to Appalachia. Even with the clear-cuts, the landscape was still beautiful, and the bubbling river we were traveling beside looked like a solid whitewater run.

After navigating several miles of narrow valley, we turned off onto the even smaller road that led to the village of Glyncorrwg. Located in a wide bowl at the head of the valley, Glyncorrwg was built around a mine and had been struggling since its closure in the 1970s. A group of local people banded together to reclaim the site for community use. The site of the old colliery is now the Glyncorrwg Ponds and Mountain Bike Centre, a world-renowned mountain bike park. We met Leigh Acteson, a local man who explained how the area has evolved:

> I'm from the village of Glyncorrwg. I'm the fourth generation to come from this community, and at the moment I am working as a project manager running the Glyncorrwg visitors center.
>
> There were mines where we are now, plus there were two mines at the top end of the valley, and they would employ over nine hundred men at their peak during the 1930s and 1940s but they closed in 1971. The Glyncorrwg Ponds are a community project that was set up in 1989 by concerned members of the community who were worried about the decline of the area. We'd already lost the coal mining, so some people came up with the idea of utilizing what we still had: beautiful scenery and lots of rain. So they came up with the idea of building some lakes. The geography of the area is basin like,

with a very steep access road. Other companies and projects tried to come in but found it difficult to maintain. So, a lot of nothing happened until we started doing this ourselves.

We went out and got the support of the village by selling one pound shares in the project and that raised two thousand pounds. We went to the Welsh office then and said, "We've got this project and our whole community is behind it. Can you help us out?" The Welsh office had the Valleys Initiative in the mid-eighties. They gave half a million pounds for five projects in Wales, and we were fortunate enough to be one of those projects. They liked our idea. It was so out there for a village in Wales to be thinking about developing lakes and developing tourism where it was initially a coal mining area.

Then, we developed a caravan park and little community businesses. The biggest project we did in 2006 was develop a mountain bike centre. That brings in over sixty thousand visitors to the upper Afan Valley and to our community every year. We really worked in partnership with the Forestry Commission to

Mountain bikers prepare to ride the trails on the former mine site in Glyncorrwg.

develop the mountain biking trails. They helped us with our initial bid that got us working towards the trails and facilities we have today. One magazine even voted our trails in the top ten in the world. Our little valley is up there with the Himalayas, the Alps, and other parts of the world.

From when we started, there were one or two bed-and-breakfasts in the area. Now there are fifteen or twenty. At our center we employ twenty-five local people who otherwise might not find work in their own valley. We can do all this. But, the regeneration of the community itself and the people within that community to see the benefits of what they've

The road to Glyncorrwg.

got—that's the longest and hardest goal that we've still got to achieve.

PEN Y CYMOEDD WIND FARM

While talking with Leigh, we learned that a Swedish energy corporation named Vattenfall was planning to build a large-scale wind farm on the ridges above Glyncorrwg. He explained that concerns voiced by local residents about the scale of the project led the company to reduce their proposal from two hundred and twenty turbines to eighty-six. Still, the Pen Y Cymoedd (which translates to "Head of the Valleys") wind project would be the largest land-based wind farm in the UK.

At first I was shocked by the opposition to the wind farm. In Appalachia I had worked on a campaign to support the Coal River Mountain Wind Project, composed of residents of the Coal River Valley in West Virginia who were proposing to build a wind farm on

Leigh Acteson is interviewed for the After Coal *documentary.*

a ridge threatened by mountaintop-removal coal mining. Although support for the coal industry, especially deep mining, remained strong in Appalachia, most Appalachian coalfield residents seemed to support wind farms over mountaintop removal as an energy source. Our mountain people saw wind farms as a way to keep their mountaintops intact and streams from being destroyed.

And yet, in Wales, many local people were against a similar proposal. It took several interviews and informal conversations to figure out why, and I am still not sure I have it right. I believe much of the opposition was aesthetic. In the US, I have seen similar concerns in places such as western North Carolina, where people have fought to keep ridgetops intact to support the unspoiled mountain views expected for retirement and vacation homes. In Wales, many of the people I spoke with felt a similar attachment to their landscape. Perhaps a more important factor was who would benefit from this development. Although the wind farm was to be built on public land (much of it old National Coal Board land now managed by the

View of wind farm from the Afan Valley.

PERSONAL HISTORY: STEVEN HOLDROYD

Steven Holdroyd is the project manager of the Pen Y Cymoedd Wind Farm. We interviewed him at the construction site, where he explained how he had made the transition from working in coal to a career in renewable energy:

I moved to Wales in 1984. I'd been appointed as a shift manager at a local coal briquetting works. At the time the coal industry was king here in south Wales. You couldn't avoid it, wherever you went. There were blue and yellow signs with National Coal Board written on them. There were fleets of buses ferrying people to and from different collieries. So it was very much an industry of the entire community at that time. And not just the people who worked down the mines. The people supported the

mines. So did the engineering companies, and the transport who got people to work, and many others. It was a thriving industry.

There were forty collieries operating deep mines in south Wales in 1984. Today there is one. That is the main difference from then to today. In 1984 the coal industry was in full swing. People lived locally and people worked locally. You could literally get up, get dressed, and walk to work. That situation has changed completely in the last twenty years. People now have to travel quite long distances to work. The main focus of people's employment is in the city centers.

To address that in certain respects, there have been a lot of investments in road links to the city, dual carriageways, and very new roads from the city up into upper reaches of the valleys where the coal mines used to be. There's also been the reopening of rail lines to passenger traffic and new stations built, all to enable people to get to work from the former mining communities to where the work now is.

Which is based in the city. So it has been a major change.

There's also been a legacy of unemployment. The old generation who were coming to the end of their careers was able to take early retirement and kept their sense of work. But there has also unfortunately been a small part of people who haven't worked since the collieries closed. There are some families in the areas which are now into a second and sometimes third generation who've never actually had a job.

Towards the end of my time in the coal industry, I was a manager. Then the plant closed and I had to look for other things to do. I was able to use some of my experience in gas extraction, and power generation from that gas extraction. That developed from just a pure gas company into a renewable energy company that took on the management of small-scale hydroelectricity assets and also started to develop wind farms. I'd been managing those assets for four or five years and then I was asked and joined this company to develop this project.

Above: *Stephen Holdroyd interviewed on the site of the Pen Y Cymoedd Wind Farm.*

Councilor Scott Jones explains the location of the wind farm to the After Coal crew.

Forestry Commission), a private company from Sweden would own the wind farm and profit from the electricity produced. In addition, most of that electricity was destined for relatively wealthy and populated towns in the south of England, leaving little to no benefit for communities in south Wales.

To better understand local opposition to the wind farm, I spoke with a local government official, Councilor Scott Jones, and asked him what people in his constituency were saying about the project. He reported:

> The majority of local people were against them [wind farms]. Our focus is to keep local people in employment. Wind farms may be, in some people's eyes, a way forward in terms of climate change, but it doesn't help the position in terms of unemployment or sickness. It might help the environment, but it doesn't help the local people and their problems.
>
> The only thing we have got is tourism and before long we're not going to have that. There's wind farms being put up left, right, and center, and I don't think that appeals to

people in terms of tourism. The project has been approved by a minister in London. So, if we are going to have them, we need to make sure that the communities that are going to have the devastation are going to receive the best money, and that the money doesn't filter into other places. If we're going to be the worst community that's going to be impacted by these wind farms then we would need compensation, as I would say, for putting up with them.

To learn more, I reached out to the Vattenfall, who put me in touch with Rahel Jones (no relation to Councilor Scott Jones), a local woman who was in charge of community relations for the Pen Y Cymoedd project. She explained why she chose to work for the wind farm:

I'm not particularly old, but when I was little I remember some of the coal mines still operating in the area. Even now that feels like quite a long time ago; the change has been massive. I've done a number of things from teaching to working in politics. But the job I had before this was in sustainable development. I focused on what is sustainable development, and what does it mean for health, education, etc. . . . I sort of got fed up with just talking about the need to be sustainable, and the chance came to work on a project in my area that was exactly about sustainable energy. It forced me to think about the needs of people in the future and really transforming our energy system and communities along the way.

With the wind farm, you get a lot of random concerns. When you talk to people you have to really listen and figure out actually what their concern is, not the dozens of concerns they've thought of because they feel the need to justify their position. The concern generally all comes down to the visual impact. Some people don't like looking at them. Some people do. But, if you don't like looking at them and you find out

Rahel Jones shows the After Coal *crew the site of the Pen Y Cymoedd Wind Farm.*

you're going to be looking at them from your community, you want to minimize that as much as possible.

And I would say that on this project we were quite lucky with the design because of the nature of the land. You've got quite a steep ridge up to the wind farm and, if you think in terms of line of sight, just moving the turbines back a little bit you can actually minimize quite a lot of the impact on communities. So we were able to do a lot of things like that to address local concerns.

If you went around the communities now, we're not pretending that everyone there is saying: "Oh, we really want this wind farm." It's a long-term project, and it's our job to show people benefits over the long term. Five, ten, or fifteen years down the line, we want people to say: "I can see that this project has brought us these opportunities." It's a long game, instead of trying to convince people to support projects like this overnight.

COMMUNITY FUNDS

The story of the Pen Y Cymoedd Wind Farm introduced me to the concept of community funds. Talking to people in Wales I discovered that when a private company in the UK undertakes a large project that disrupts a local community, it is common for the government to require the company to create a fund that helps the community adapt to the disruption. Usually, these funds are distributed by a group of local residents, which is a key democratic principle. Although this is a standard practice for any large infrastructure project in the UK, it is unheard of in Appalachia, where local politicians usually bend over backward to offer tax incentives for private companies to extract resources.

When we interviewed Rahel Jones, she explained how the fund for communities affected by the wind farm will work:

> The big part of our project for the communities is the community fund. So we've committed to a community fund that's worth about 1.8 million [pounds] every year. And for the area it's guaranteed for the life of the project. Obviously having 1.8 million every year and knowing it is coming every year for twenty-five years is a pretty big deal and doesn't come along every day. Having said that, doing it well is quite difficult.
>
> We're spending a lot of time at the moment with the communities, taking a step back and asking what is needed for these communities to survive in a sustainable future. So we've been doing some research, we've been going to local shops, theatres, cinemas, and carnivals. All sorts of places, really, to talk to people about what they need for the future and can we build the fund that can deliver on those needs rather than spending it on the nice things that may be nice but won't create long-term opportunities for these areas.

TOWER COMMUNITY FUND

Another example of a community fund I learned about in Wales was still tied to the coal industry. The Tower Colliery in the upper Cynon Valley of Wales first opened in 1864. As the mines privatized in the 1990s, the National Coal Board planned to close Tower and lay off hundreds of miners. In a bold move, the miners pooled their severance pay and purchased the company, successfully operating the mine until the last of the deep coal was extracted in 2008 (Tower 2017).

Five years after closing the deep mine, Tower started an opencast mine to extract the few remaining coal seams close to the surface. Opening a surface mine, rather than a deep mine, was controversial, so the Tower miners, who still owned the property, worked with local community members to establish a reclamation plan and set up a community fund based on coal production. To learn more about how the fund works, we met Hayley Teague, the Tower Fund community benefit coordinator, in the office of the Tower opencast mine. She said:

> I'm a miner's daughter. I got to experience the year-long strike in 1984–85 because I married a miner two weeks before he went on strike. Now I work as the Tower Fund community benefit coordinator. I am in charge of a benefit, or a lump of money, that Tower donates to the local communities, and they have a separate little common fund for the rest of south Wales mining areas as well. Local people will apply and, if their projects are good and benefit the community, then we go ahead and give them the money for that project.
>
> The communities set the goals. This is the uniqueness of it. It's them that decides. The four communities are the immediate communities situated from where we are now in Tower. So, we've got Rhigos, which is just up the road from Tower;

Hayley Teague is interviewed for the After Coal documentary.

Hirwaun, which is more central; Penywaun; and Penderyn. The amount of money they get per year is decided on the population of each area. The decisions made on the applications are made by the communities themselves. Nobody—not me, not Tower, not the charity that we also use—nobody has the power to make decisions but the communities.

This year we've given out just over 223,000 pounds to the four communities and the small common fund have spent over 65,000 pounds. But we've encouraged the local communities to keep some of their money for bigger projects. The first actual legacy project we started in Rhigos, they want to expand the sports hall they've got there to provide a better gym facility that will encourage more females to use it. Another project we're doing is in Penywaun, a fantastic community. They're going to regenerate their community center. We've just started on that.

The open-cast plan is to mine over seven years, so six years are remaining. Some of these projects don't happen overnight. Getting people to realize, okay, you want a skate park or a new community center, but it's not going to happen overnight. Getting them involved from the very start brings communities together. They're all working together to solve problems and plan for the future.

Talking to people in the Welsh valleys helped me see how the legacy of coal mining could help support a more diverse post-coal economy. Repurposing old mine offices as the DOVE Workshop has done, reclaiming mine sites and turning them into mountain bike parks or similar recreation areas, and creating social enterprises and community funds were all strategies that allowed Welsh mining communities to survive the collapse of the coal industry. Certainly the area has lost population and wages are not as high as they were during the height of the coal industry, but the end of coal opened up new opportunities for groups who had been shut out of the old economy, especially women. And, most importantly, the villages survived and in many places community spirit was thriving.

Streetscapes of Blaengwynfi, Wales, and Cumberland, Kentucky.

Streetscapes of Croeserw, Wales, and Lynch, Kentucky.

Community mosaics in the Afan Valley of Wales, and Harlan County, Kentucky.

I CARRY THE COAL
I HAVE A LOT OF
CRACKS AS I DRIVE
ON THE TRACKS

CHAPTER 5

BACK IN THE USA

I returned from Wales in the summer of 2012 charged up and ready to talk with my friends and collaborators about what I had learned. I hoped that post-coal stories from Welsh mining communities would help bridge the intense political divisions in Appalachia. However, the 2012 presidential campaign was gearing up, and the political rhetoric only served to deepen divisions across the United States—especially in the coalfields where politics were intensely personal.

Pro-coal bumper sticker on a truck in Harlan, Kentucky.

Tensions around the coal industry in central Appalachia have existed for more than a century, but in 2012, massive mine layoffs and new environmental regulations created a toxic atmosphere for civil dialogue. Many coal miners and their families blamed the

Obama administration and the "tree huggers" (aka environmental-ists) for their problems. President Obama lived in Washington, D.C., but plenty of the tree huggers were local residents, and they were worried. Only two years earlier, an angry mob of coal supporters physically blocked environmentalists from testifying at a hearing on the Environmental Protection Agency's stream buffer zone rule, a controversial measure that blocks mining within a hundred feet of headwater streams in Charleston, West Virginia. Many grassroots activists were concerned for their safety. Although they were inter-ested in hearing about what I had learned in Wales, the concept of life after coal had the feel of science fiction in Appalachia.

The Environmental Protection Agency's increased scrutiny of moun-taintop-removal mining permits fed into the industry's narrative that Obama was waging a war on coal. To build support, the coal

Above: *Bumper sticker distributed by Friends of Coal ahead of the 2012 presidential election.*

industry sponsored several political rallies during the summer of 2012. I decided that I should hear what industry representatives had to say, so I drove to an event advertised as a rally for coal jobs. I followed the map provided by an organization named Friends of Coal to an old tobacco warehouse on the outskirts of Abingdon, Virginia. I parked next to a pickup with bumper stickers that read "Obama's No Jobs Zone" and "If You Don't Like Coal, Don't Use Electricity." I got out of my truck and decided to check out the scene before bringing the camera in.

Approaching the old tobacco warehouse was like entering the big top at a circus, except that the entrance was a white metal garage door set in the yellow corrugated siding of the warehouse. Just to the right of the entrance, red, white, and blue balloons floated over a small tent where a few children lined up for a chance to ride on miniature train cars. Nearby, a group of preteen boys wearing identical white shirts that read **CITIZENS FOR COAL** in bold across the shoulders and "Coalition for Mountaintop Mining" in smaller letters underneath gathered around a relic from a carnival sideshow. They took turns hitting a big metal button with a sledge-hammer, trying to send a cylinder up a vertical shaft and ring the bell on top. One kid had developed a surprisingly successful one-handed technique.

Inside, the crowd seemed to be more coal executives than coal miners. Although there were plenty of T-shirts with bold slogans, including "Legalize Coal" and "Got Electricity? Thank a Coal Miner," the rally participants were neatly dressed in khaki trousers and a surprising number of pastel polo shirts. I ducked under a few American flags and skirted an inflatable bounce house for kids, scouting the perimeter before moving down a makeshift midway at the back of the hall.

Above me colorful banners shouted pro-coal slogans:

> **AMERICA RUNS ON COAL,**
>
> **STAND WITH COAL,**
>
> **ABUNDANT COAL = AFFORDABLE ELECTRICITY,**
>
> **COAL MINING OUR FUTURE** (I had to read that one twice),
>
> **COAL FEEDS MY FAMILY**,
>
> and simply: **POWER PROGRESS COAL**.

A small crowd was laughing pleasantly around a booth toward the back, so I headed that way. As I approached, I saw that everyone was waiting for a chance to spin what looked like a cross between a roulette wheel and the Wheel of Fortune. The sections on the wheel were painted in red, white, and blue. Instead of dollar amounts, they contained a list of topics, which included Coal, Jobs, Economy, Taxes, Electricity, Steel, Affordable, Abundant, Reliable, American, Security, Fuel, and Spin Again.

When it was my turn to spin, the wheel landed on "Coal," and a young woman in a Friends of Coal shirt looked at my camera and asked me pointedly: "We only like true facts about coal, right?"

"Right," I agreed.

She looked at her card and read: "True or false: Coal's share of the international energy market is projected to continue to grow."

"True," I guessed.

"Good job!" She smiled, then waved the next person up to the wheel to inform them that the states that burn the most coal have the most affordable electricity rates.

One of the largest banners had a red map of the Appalachian region, highlighting Pennsylvania, Ohio, West Virginia, Kentucky, Virginia, and Tennessee. The text read, "The Obama administration's no jobs zone. The president talks about creating jobs, but in Appalachia his EPA is destroying jobs."

Next to it, an ominous black banner sported large white block letters that read **NOVEMBER IS COMING, AMERICANS FOR PROSPERITY**. To be honest, I did not pay much attention to that sign at the time. It was not until I was editing footage from the rally that I realized that Americans for Prosperity is the political action committee created by billionaire brothers Charles and David Koch. The Center for Public Integrity reports that this Astroturf group spent $122 million backing Republican candidates in 2012 (Beckel 2013).

Americans for Prosperity and other coal industry–funded political action committees were pouring record amounts of money into the 2012 political campaigns. Most of these funds were supporting the Republican candidate for president, Mitt Romney, but some were targeted for state-level races. As a result, many Democrats from coal-mining states bent over backward to prove that they could also be staunch supporters of coal. State representative Rocky Adkins, a Democrat from Sandy Hook, Kentucky, was the house majority leader at the time. He took the stage shortly after I entered the warehouse and turned the volume up to eleven:

> I am PROUD to have a coal-fired power plant in my district. I want to talk about this coal-fired power plant here for just for a minute, because the coal-fired power plant is under attack.
>
> They are saying that they want to shut that coal-fired power plant down and bring power in from other parts of this country, from other states, and produce power for our people in eastern Kentucky. Are you going to tell me that you can produce power cheaper from another state than you can produce from that coal-fired power plant and from that natural resource of coal that is right under our feet?
>
> Tell American Electric Power to stand up and fight the

federal EPA! We're not going to shut that coal-fired power plant down! We're going to continue to mine coal, and we are going to continue to burn coal! We cannot allow AEP Kentucky Power to wave the white flag of defeat in the heart of coal country and allow the federal EPA to declare victory in the middle of the commonwealth of Kentucky, in the middle of coal country. We need to stand strong, we need to stand proud, and we need to continue to mine coal and produce the energy that this country needs today and is going to need in the future. Are you going to stand with me as we fight this battle together?[3]

The crowd applauded madly as Rocky shouted, "Do we have any coal miners in this crowd? Do we have any friends of coal in this crowd?"

As the applause died down, he continued:

Let me tell you something, there is a war on coal, an attack on this industry . . . Everyone in this room is all for protecting the environment, we're all for that. Folks, I tell you what we're doing today, we're mining coal safer and better than we ever have in the history of time. But more importantly, we are burning coal cleaner today in the United States of America than we ever have in the history of time.

Why bring these regulations that impact our economy and put our people out of work? Use common sense! Allow us to continue to mine our coal! Allow our people in central Appalachia to continue to work and have a good job!

Politicians throughout the coal states echoed these themes in their speeches, blaming the decline of the coal industry on unnecessary government regulations. Meanwhile, industry analysts were telling investors that competition from cheap natural gas (produced by the controversial fracking process) was the largest factor in coal's decline (Culver and Mingguo 2016).

DIVIDED WE FALL

The political rhetoric around coal in Appalachia left little room for dialogue about diversifying the economy. More importantly, the sound bytes served up on TV news and social media left no room for politicians or local residents to discuss job training, mine reclamation, community funds, or other ideas to support miners and industrial workers who had lost their jobs. During this campaign season, I sensed that many Appalachians were not ready to talk about life after coal, fearing that the idea would be perceived as anti-coal and deepen divisions in their communities. This puzzled me, but I was resolved to meet people where they were, not where I wanted them to be.

Reflecting on my recent travels, I remembered that many people in the Welsh valleys still resented the Conservative government for shutting down the nationalized coal industry in the 1980s. However, Welsh labor advocates were still engaged in the political process, pushing the government to play an active role in regenerating their communities. In contrast, many Appalachian coalfield residents blamed the liberal government for shutting down the coal industry, and viewed the government and all of its programs as the enemy. As a result, they disengaged from the political process and ceded power to wealthy corporations.

In Appalachia, the struggle for the After Coal project was how to move past the rhetoric of pro-coal versus anti-coal toward a conversation about what is good for the community. Retired miners like Rutland Melton felt torn by these deep divisions. He supported underground mining, but was staunchly opposed to mountain-top-removal coal mining and resented industry attempts to describe his position as anti-coal:

> They think we're trying to stop mining coal. We're not trying to stop mining coal. Do it the way it's supposed to be done. You know, the right way.
>
> I don't have nothing against mining coal because I was in the mines twenty-three years, but I am against that mountain-top removal. That's just a cheap way of mining, working with less men. I look at it this way: God created the mountains to be admired, not to be destroyed. You can't make those mountains look the same as they did after you tore it up.

Lauren Adams, an eighteen-year-old with dyed blue hair who had grown up just down the road from Rutland, also felt torn. She was from a mining family and was concerned about a future without the industry that had created her hometown of Lynch, Kentucky:

> Around here, coal mining is a big part in a lot of people's lives. For so many families, that's how they put food on the table. It's extremely important to them because it is their job, that's how they are putting clothes on their kids and stuff. And then anti-coal people will bring up things about the environment and how it might ruin the air and water. I guess they are trying to think about how coal isn't going to last forever, because coal miners are always getting laid off left and right.
>
> I guess I could sympathize with both sides, I don't want to destroy the environment, but I also want to feed my kids. A job is a job, and there's not many jobs around here. I knew a lot of people who dropped out of high school and became coal miners to support their families. It's really either join the mines or move away.

Lauren Adams on the stage of Higher Ground 4: Foglights.

TALKING THROUGH CHANGE

Making space for conversations about what is best for coalfield communities was the reason I decided to make the *After Coal* documentary, so I sought out the places where these discussions were already happening. My goal was to amplify local discussions in order to increase support for community-led efforts to diversify the Appalachian economy. My next stop was the campus of Southeast Kentucky Community and Technical College in Cumberland, Kentucky. Robert Gipe, director of the college's Appalachian Center, had been working with community members (including Lauren Adams and Rutland Melton) on a series of art projects designed to start conversations that address local issues.

The organization, called Higher Ground of Harlan County, was launched in about 2001 in response to the prescription drug crisis (which predated the current opioid epidemic). Higher Ground's multi-generational and multiethnic participants gathered local stories, developed participatory photography and mural projects, and eventually created a series of original plays. The project had gained national attention, including grant awards from the National Endowment for the Arts and ArtPlace America. In 2013, they were starting work on a new play about the future of Harlan County. The idea was that conversations around the scripting, rehearsals, and performance of the play could help local residents define their own future.

Lauren Adams told me about her experience participating in Higher Ground:

> You can't talk about stuff around here. It's really hard to talk about anything: coal or different religions, or different sexualities. You just can't. Our play was just miraculous to me, because everyone loved it and it talked about coal. The play made a situation where it was safe to talk about it. You saw a fictional family that honestly was every family, you know. It was safe to talk about and it went over really, really good.

This was one of the few cases where people had figured out how to reach across the deep divides in coalfield communities, so I decided to spend time documenting the production of *Higher Ground 4: Foglights*. In order to reach the broadest possible audience, this play would be performed off campus, in four historic sites throughout Harlan County during the fall of 2013. The title *Foglights* had caught my eye, so I asked Robert Gipe to explain how they had decided on the name. He said:

> The lighting designer visited Harlan County and he was so struck visually by the lay of the fog on the mountains. We were like, "Yeah, I guess it is kind of cool." It's one of those things that are so much a part of a place, and even though you love it, you don't think to remark on it. Then it struck us that the fog is a great metaphor for right now. Because the bottom

was dropping out of the coal industry unlike it had ever been, and it seemed like everybody was talking about how foggy the future was.

DOCUMENTING FOGLIGHTS

My first shoot was a ride along with Robert Gipe as he took theatrical director Richard Owen Geer (no relation to the movie star, who spells his last name differently) on a tour of the historic sites where the performance would be held. Many of these sites did not have a stage, some did not even have a theater. One of our first stops was a sawmill, where Robert negotiated for local lumber to build a stage that would travel with the show to each location.

Throughout the summer of 2013 I followed Robert, Lauren, Rutland, and many others as they pulled together to make this traveling show. Higher Ground participants ranged from high school and community college students to retired miners and ministers. Many of the summer classes taught at the community college contributed to the production. Students recorded oral histories, collected old photographs, and learned a DIY process to transfer photographs to fabric. They used this technique to create giant banners that would welcome the audience into the performance spaces.

When I recorded this group, Lauren Adams was helping organize their work. She demonstrated the iron-on process that transferred the student's images from computer paper onto the banners, then introduced me to her friend Candy, who had enlarged an old photograph of her grandparents. The photo depicted a handsome middle-aged white couple standing in front of a green hillside. Candy was in the process of using an iron to transfer pieces of the image from paper to a banner.

The "Deadeye" banner, created by Candy Miller, toured with Higher Ground 4: Foglights. *It was also included in an exhibit titled* By the People: Designing a Better America, *displayed at the Cooper Hewitt, Smithsonian Design Museum, in New York City between September 30, 2016, and February 26, 2017.*

"Did you tell this dude about Deadeye?" someone asked her.

I had no idea what they were talking about, or even if the question was addressed to me, so I just shook my head and started to set up my next shot. Lauren poked me. "Deadeye was her mamaw," she said meaningfully.

Then I understood that this was the time to ask the obvious question: "Why did they call her Deadeye?"

Candy explained that when she was a little girl, a neighbor had spotted a copperhead in her grandparents' driveway. Hearing the commotion, her mamaw came outside holding a pistol. Standing on the porch, she took aim at the viper stretched across the warm gravel some fifty yards away. With one eye closed, she breathed, pulled the trigger, and fired a single shot. The snake flipped in the air, its head a bloody pulp. "One shot was all she needed," Candy confirmed. "After that everyone called her Deadeye."

In downtown Cumberland, an art class was busy making backdrops out of junk. The college had recently acquired the abandoned Rowlett Furniture store on Main Street. Their long-term plan was to renovate the space into an arts education center, but the first task was to clean out the old junk. Students gathered the most interesting trash and piled it into the old showroom to develop a set of backdrops for the play. Another student had brought materials from an old coal truck garage to add to the mix. Lauren explained how the backdrops connected to the content of the play:

> A big part of the new play is about junk. There's a character who has a hoarder for a father and she really wants to live near him, but he won't move all of his stuff, so she has to leave. It's like she can't get through to him and it's just kind of the theme throughout the play, all of this junk. That's why we used a bunch of junk that we found in the Rowlett building to make these backdrops.

I asked her: "Do you think moving junk is a good metaphor either for yourself, or for the future of Harlan County?"

Lauren answered: "Yeah, ever since we started writing this script it's made me just kind of think about the junk in my life. I've dealt with some stuff, and so has everybody else, because it makes you think about the junk we need to get rid of. It also helps me imagine what we can make out of all the old junk in our lives."

Stage backdrops for Higher Ground 4: Foglights *made from salvaged material.*

Higher Ground musical director Ann Schertz speaks to actors before the opening of Higher Ground 4: Foglights.

The junk also served as a reminder that the old industrial mindsets that hold the region back need to be cleaned out to make space and opportunity for young people to create their own future. Eventually the students rearranged the letters over the old furniture store. Today, the letters "Rowlett Furniture" spell out "Future Town."

In addition to writing a script, building a stage, and making backdrops for their production, a band and a choir formed to create an original score for the play. It was a dynamic setting, with creative energy that gave me hope for the Appalachian coalfields. Ann Schertz, the musical director of Higher Ground, explained how the musical elements would help unify both performers and audience, supporting the goals of Higher Ground:

> The new play is a lot about intergenerational things. How the future is foggy and how do we decide how to carry on in a new way. You know, we have situations that call for new ways of thinking, and it's not easy to see the way. It is especially not easy for young people to see how their future is going to be, given the past in this community. A big part of it of course is coal, and how are we going to figure out what to do if coal isn't the biggest

thing in the economy. But it's really also about intergenerational things.

We want to be able to use these plays to increase dialogue, you know. How do we bring people together across barriers that have been in this community for so long to talk about things that are not easy to talk about? It does represent something about the current situation where the Right doesn't talk to the Left, and the polarization, and not being able to get anything done on the bigger stage. You know it's very [much] the same in this community. How can we use the plays to encourage people to not give up on the people that are different, that think differently from them, and keep this conversation going? I think that's why these plays have been important.

Higher Ground comes from a coal-mining tradition, among other traditions, but let's use coal mining as a metaphor. Higher Ground is a mining operation and it digs deep into these hills and into the hearts of the people to find something of value that can be brought to a market. It isn't a market that's turning on the lights in Chicago, it's a market that is brightening the eyes, raising the horizons of the people right here.

What people find here is like coal, stories laid down from the strata of other times. By excavating those stories they find a wisdom that is energy, and that energy has the ability to make a difference, to be a kind of light, and with that everything else can begin.

Stories alone don't make it happen. But, unless people connect to the sense that this is the center of the world, until people understand that this place is where I can make a difference, then nothing else can happen. You can't start a house-building program, you can't start an education program, you can't start anything unless the people believe they can make it better. That's the vision without which the people perish, and that's what comes from Higher Ground.

—Richard Owen Geer, theatrical director
of *Higher Ground 4: Foglights*

Theatrical director Richard Owen Geer during his interview for After Coal.

During the final weekend of the month-long run of *Higher Ground 4: Foglights*, I interviewed Robert Gipe about the experience. So many things had happened during the production of the play that I started by asking him to describe the Higher Ground project in the simplest terms possible. He said:

> Higher Ground is art centered, but it's really about the people in this community. Higher Ground is Harlan County. Higher Ground is about how do you do something that brings people back to the idea of the common goodness in this place.
>
> Higher Ground of course is a theater project, Higher Ground is a story-telling project, Higher Ground is helping people in this community decide how the community looks in visual arts, and Higher Ground is music, music performed together. I love what Ann always says: "It's not about singing louder, it's about hearing each other and singing together." The other thing she said was: "It's easy to let the people who are the strongest carry the music, but it sounds way better if you all sing." That is what it is all about, right? You can let other people do it. But, if you join in it sounds better. It's easier for people who have a hard time

Higher Ground project director Robert Gipe during his interview for the After Coal *documentary.*

singing to not stand out. I think there is something profound in that, if we just will all do it, nobody is going to sound bad. Everybody is in it together.

What you do within the theater space is a super-powerful metaphor for what you hope to accomplish in the community.

—Robert Gipe, director of the Appalachian Center at Southeast Kentucky Community and Technical College

I think that a big part of what is important about Higher Ground is the continued making of new things. We're continually rewriting what the place is, we're continually creating. It's hard, but the fact that we continue to create things becomes evidence that it doesn't have to be the way it's always been.

The big question is: Is this community capable of change? Could this community be something different than what it is? If you went down and asked people on the street "Is this community capable of change?" there would be a substantial

The cast of Higher Ground 4: Foglights.

number of people who would say, "No, no it's not. It's always been this way, always will be this way, that'll never happen." These are some of the comments we've heard forever doing this. So we started reframing the script around these questions of "Does it have to be this way? Do we have to have the same junk in the house all the time? Is there going to be a place for me? Can I be who I am and still be here?"

All of these questions have to do with people's beliefs and what they're willing to include in their definition of themselves or their place. One of the things that evolved is that the idea is that all we can be is a coal-mining community. One student wrote in her summer essay: Is that all we are? Aren't we more than that? I think that that idea that we are this way, that we can only be this way, and if that fails then we're done is very depressing and very difficult for a young person to swallow. Then this whole question of "Do we have the capacity to change?" becomes the question that enfolds all kinds of other questions about this moment.

Robert Gipe continues on p. 142

Robert Gipe (right) *with Carrie Brunk* (center) *and Rutland Melton* (left) *at the Eastern Kentucky Social Club.*

I can't get out of here fast enough. The buildings are falling apart and so are the people. All anyone cares about is coal. Out of all the things we are, we pick coal? I appreciate every miner that's ever been. But I am more than a rock in the ground, and so is this place.

—Lauren Adams, a teenage participant in
Higher Ground 4: Foglights

Robert Gipe, continued from p. 141

You have to show the community that anything can happen. The unexpected quality of Higher Ground is important. The sense that, if you can pull this off, then it becomes a living example of how meeting a challenge is good for you. It gives everybody a sense of their own capacity. The tour grew out of this idea of how do you help use the arts to infuse some kind of spirit into different places. I felt like that it was important to push it, to push the play out.

I mean it's really quite a journey, we started in the town of Lynch, in a building that was built in the early 1920s as a segregated school. It was built by the biggest corporation in America, United States Steel, to educate its black children. Teachers at the school had PhDs and they were teaching high school because they couldn't get jobs because of segregation and racial discrimination. The school had been saved from destruction by the Eastern Kentucky Social Club, which is this amazing organization that held together a community of black people who were Kentucky coal-mining families, even as they spread out all over the world.

Then we went to the old Evarts High School, which is on the Clover Fork. It is another high school that had meant the world to the community. The community center there is where all the community meetings during the Brookside strike were, and it's where the Save Black Mountain hearings were. The old community center had gotten destroyed as part of a flood control plan and the group that ran the center received money from the Corps of Engineers to save the old high school building. They're trying to do something with it so they invited us to do the play there.

One of the things that made this tour so difficult was just the infrastructure problems with the old buildings. These sites are maintained by volunteers, so they're battling plumbing problems and they're battling keeping the doors open. We rehearsed all summer without air conditioning; it was just hot. That affects how many people come to practice and that affects how complicated the choreography can be. Then, all of the copper had been stolen out of Evarts. Our electrical technology teacher is an unsung hero; he went in and rewired two of our sites.

In between those two places we were at Pine Mountain

Settlement School, which has been there since 1913. We didn't try to compete with the natural beauty there, so we performed under a tent. That was its own unique experience, it was like a revival. It was so intimate, we pushed all the chairs close together so that it was like you were sitting in this big living room.

For the last show we're going to be in the Harlan elementary school auditorium. It's more of a modern facility, but it's on the site of the original Harlan High School gym. The old school burnt, so when they rebuilt this gym they actually used the original steel frame of the old gym. It's just like that story in the play about the guy burning down the log house to get the nails. You can use the old to build the new.

The places that we've been to on this tour were picked out of necessity. But it makes a kind of capstone for this story that's exactly what part of the message of the play is: You can build new without having to abandon the old.

COMMUNITY FEEDBACK

After documenting the Higher Ground project, I pieced together a rough draft of the *After Coal* film that placed the stories of Wales and Appalachia side by side. Each time I screened this work-in-progress documentary, a lively discussion ensued, which told me I was on the right track. However, several people pointed out to me that the Appalachian stories were more about the arts than economic development. As a person who believes that arts are an essential component of economic development, this presented a challenge to me. Clearly, the first draft of the documentary was not demonstrating the role of arts and education in creating sustainable jobs.

A lot of people wanted a concrete answer to the question: Where are the jobs if coal is gone? I felt that this was a legitimate request. Even though I did not have a clear answer handy, I believe that coalfield residents deserve straight answers about their future.

It was not until I shared these concerns with project advisors Hywel Francis and Helen Lewis during Helen's ninetieth birthday party that I realized one of the answers had been right under my nose. They told me: "If you are looking for long-lasting businesses in the coalfields that pay a living wage, then Appalshop needs to be a part of the story."

"But I can't do it, I'm too close," I protested to my advisors. (Full disclosure: I worked at Appalshop from 1990 to 2007 and still keep close ties with the organization.) Try and see what happens they recommended, so I did. I realized they had a point. Appalshop has survived almost fifty years, longer than any coal company in the communities it serves.

The organization started in 1969 as the Appalachian Community Film Workshop of Whitesburg, Kentucky, a federal war-on-poverty program to train disadvantaged youth for jobs in the film and television industry. However, the young trainees at Appalshop decided that their new skills would be best used to preserve Appalachian culture and protect mountain communities, so they created their own non-profit and shortened the name from the Appalachian Community Film Workshop to Appalshop. The organization currently employs more than twenty people, and Appalshop's education and training programs have provided job skills to several generations of coalfield residents.

Appalshop's work is built upon the belief that "communities flourish where there is a deep and abiding culture that draws on strong traditions and creates new traditions," and that "people have the right to control the development of their own cultures and communities, and have a voice in public life" (Appalshop 1994). Today, a new generation of leaders is moving the organization into its fifth decade. I spoke with Alex Gibson, executive director, and Ada Smith, director of development, about the organization's role in regenerating coalfield communities. Alex Gibson said:

> The industry at the center of any small town does a great deal of work, indirectly supporting businesses such as retail, transportation, or light manufacturing. Here in Whitesburg, coal companies were central to our economy. As the economy changed, coal and associated retail industries started to suffer because they were linked to coal prices and the cost of extracting this particular mineral from the earth.
>
> Appalshop has remained after the coal mines closed. Today, one of the most enduring industries of this city is our art industry. Appalshop's work has led to a lot of spin-offs. There are several businesses up and down Whitesburg's Main Street run by former Appalshop employees, shops like tattoo parlors, restaurants, or farmer's markets. Then, there are companies that we directly incubate. Mountain Tech Media is an example of a software company that currently employs two people, but we expect that that number will rise substantially in the future. This year (2016) is their first year of operation, and they will do a hundred thousand dollars in revenue. That's a number that doesn't really get competed with in this area in terms of a startup business.

Ada Smith said:

> We have an impact on this county of about two million dollars a year. Most of the salaries we pay get re-spent locally. Maybe, in the twentieth century, people felt like arts and culture was touchy-feely and not important to our economy, but I think

Appalshop executive director Alex Gibson (left) *and director of institutional development Ada Smith* (right).

in the twenty-first century, it is essential to employment. I think that for hard economic development to happen in a twenty-first-century economy, you have to have a sense of place—of what unique assets you can sell. Obviously, arts and culture have a huge role in helping to create that community identity and sense of place.

Appalshop began as a training program and continues to do a lot of training for young people and people of all ages. For example, many people who start out as volunteers at our radio station go on to careers in radio and television, including Ryan Adams, one of our local newscasters. Another example is our Appalachian Media Institute (AMI), which uses media production to develop leadership skills in young people for our region. One of our first trainees, Nema Brewer, is now the head of communications for the Fayette County Public School System.

Then there are people like one of my good friends, Brittany Hunsaker, who went through the AMI program and she's now a social worker at a local health care facility, Kentucky River Community Care. She doesn't work with media, but her whole understanding of this place and how to work with people to solve problems came from her experiences at AMI. So, there are tons of examples. Many people we trained now work in media or the arts, but just as many people don't have a job that directly relates to what Appalshop does, but their process of understanding this place and themselves happened here.

Finally, I think one of the things Appalshop has done over the past forty-seven years is create an environment where trial and error is accepted, understood, and a part of daily life here. More and more people aren't scared to try something different, and even if they are scared, they understand there's some people around them that are also doing weird and different things, and that it's not such a foreign concept.

We have to be able to imagine and see things that aren't in front of us. Nobody expected the internet to be what it is. Nobody could see that and yet people imagined something different. And now it's here. So I think art and culture is touchy-feely in a good way, because you have to have a strong community identity and you have to connect people to each other in order to build something together and create real economic opportunity.

REGENERATION THROUGH EDUCATION AND AFFORDABLE HOUSING

Although Appalshop's story explains how the arts help build a sustainable economy, a sustainable economy needs more than just arts. My next stop was the Southern Appalachian Labor School (SALS). SALS has a long connection with the After Coal project. During the 1970s, SALS cofounder David Greene participated in the exchange that brought Welsh miners to Appalachia, and in 2012, SALS served as a fiscal sponsor for a grant from the West Virginia Humanities Council to support the production of the *After Coal* documentary.

The Southern Appalachian Labor School was founded in 1977 at West Virginia Institute of Technology in Fayette County, the heart of the historic New River coalfield. The organization was created to provide continuing education opportunities to union coal miners, industrial workers, and other underserved populations. As the coal industry declined, SALS kept its focus on education but adapted its programs to meet the changing needs of the communities it serves. The school's mission is to "provide education, research, and linkages for working class and disenfranchised peoples in order to promote understanding, empowerment, and change" (SALS 2016).

Today, SALS runs a range of programs, many of which initially appear to be infrastructure projects. However, SALS integrates education and training into all its projects in order to increase the capacity of the communities it serves.

In 2014, graduate student Cheryl Laws and I interviewed several SALS staff to create a profile of the organization. We talked to John David, one of the cofounders of the Southern Appalachian Labor School who currently serves as the organization's director; Kathryn South, the organization's YouthBuild project director; and Vickie Smith, SALS construction manager.

We met them at the Historic Oak Hill School in Oak Hill, West Virginia. This impressive three-story brick structure was built in the early 1900s. After sitting abandoned for many years, SALS purchased it and began renovation. As part of an effort to regenerate the community, they are transforming the former school into space for small businesses, training, and housing for veterans. The grounds host a community garden, and SALS operates a low-power

One of many homes that the Southern Appalachian Labor School has renovated in Oak Hill, West Virginia.

radio station (WAGE—106.5 FM) as part of their community regeneration efforts.

Vickie Smith provided more details:

> We're just two blocks off of Main Street. Our dream is to take the first floor and make it an enterprise where people can start up businesses, small businesses, locally, at a fraction of what it would cost to do so elsewhere. We would like to house the veterans, or homeless, on the third floor.

John David added:

> We're also working to obtain tax incentives for folks through the low-income tax credit program and the historic tax credit program so investors will get a tax break if they work with us to help improve the facility.

Sitting on a bench outside the school, John David reflected on how the organization has adapted to the changing needs of area communities:

SALS was founded to provide worker education to folks who are underserved. West Virginia has always been a wage state, and the union jobs provided spending money for other businesses. That is how the state operated for a long time. Now, with the union jobs in chemical, steel, glass, and coal disappearing, the economy itself has shrunk.

A big piece of our programming is designed to deal with the fact that many of the folks, particularly young adults, have dropped out of school for various reasons. We have focused a lot on working with youth that need a GED or a high school diploma and other kinds of training that we can provide to move them forward. The government classifies them as "at risk," but we believe they are "at promise."

Barb Painter runs the green building program. She took us to a house that several trainees in the GED program were in the process of renovating. As we watched young men replace the floor, she explained the connection between education and housing renovation:

As they're getting their GED they also go through a series of classes with us here at the school. They get certified in weatherization and green building, they get certified in first aid or building maintenance. Everything we do here at SALS has some kind of training extension to it. Once young people are trained in energy efficiency, they have a marketable trade, they have experience, which every employer wants, and they can go out and find jobs with local agencies. Private contractors are also more into green building now. When youth are able to find a job without relocating, it puts the money back in the local economy.

Vickie Smith explains why there is need for affordable and safe housing in a region with a relatively high rate of home ownership:

Many of these houses were built really quick, they were not supposed to last a hundred years. They were supposed to last for five, ten years and be done with. But, our economy never moved forward enough so that people could afford to move from that homestead. Now you have generation after generation that inherits a home, and that's all they have, that's their stability, they're not going to leave it. Our goal was to bring that up to today's standards.

We've went in and found snakes in the walls, dirt floors, wiring wrapped around the house, no heat system, a coal stove in the middle of the floor. But our grant support allows us to go in and upgrade the home and make people safer, healthier, and happier. If a child doesn't sleep very well because rodents are running all over the house, or there's no heat, or there's not adequate lighting, then they don't do well in school either, because they're tired. They might have health issues because of it. So, housing has a lot to do with the well-being of the people in our community.

It's really full circle because once people come into the historic school, they meet our youth, there's relationships built, and people are more apt to go out and help their community once they feel safe and secure. We get them into the feeding programs that we offer and the education programs that we offer, and then the training programs connect youth to our community. I mean it helps everybody in the community really, not just those that need the help. I mean, as you upgrade homes and make their value better, the value of everybody's homes goes up.

John David connects SALS programs to a spirit of renovation:

We build relationships in community. It creates a sense of pride, a sense that there is something going on. In terms of tomorrow, of the future, we're talking about transition. What are we going to do after coal, what are we going to do about the economy that is transitioning into something else?

Can we help folks, themselves, set up social enterprises? Can we help people, themselves, set up a Time Dollars program? This would be a new currency where their time is respected as value, and everybody is appreciated for what they can contribute, regardless if they have a degree. In terms of community, coal camps were all once thriving places. We can still come together as SALS and be a viable economic entity to do good things for folks, because folks want to live here. This is their home, this is their place.

LOCAL FOOD AND LOCAL ECONOMY

The Southern Appalachian Labor School is a great example of using local resources to meet local needs and provide vision and opportunities for the future. Another example is the Grow Appalachia project. The program, run out of Berea College, serves communities across the Appalachian region from Ohio to Tennessee and works to promote food security, nutrition, and sustainable farming practices in many former mining communities. I learned about their work in the coalfields from Valerie Ison Horn, the project director of Letcher County Farmer's Market and an old friend. She is running programs that support local agriculture and provide healthy food to people in need. Valerie connected me to Shane Lucas, one of several former miners who have become farmers that are connected to Grow Appalachia. On a warm June afternoon, I visited Lucas Farm to interview Shane for the *After Coal* documentary.

I pulled into the driveway just as he was selling a basket of tomatoes to an elderly couple. I waited for them to finish swapping garden tips then introduced myself. Despite the agricultural setting, Shane made clear that he still identified with the industrial work of coal mining.

SHANE LUCAS: Coal mining was very good money, I hate to give it up. Two years ago I had the best job I ever had in my life. Probably the best job I'll ever have in my life. Probably the best job that you go anywhere in the United States and get. And it went down.

TOM HANSELL: That quick?

SHANE LUCAS: About two years before that, I knew it was coming to an end, so I started selling vegetables on the side and thinking about going into farming.

TOM HANSELL: How many years did you end up working in the mines?

SHANE LUCAS: Seventeen or eighteen years. I was running heavy equipment. I'm a drill operator.

I knew that there were going to be layoffs, you know coal was getting bad. I knew that I had to do something to make a dollar, so I decided I want to do vegetables. I started selling on the side of the road. Then I got a little bit bigger and I said, well, it's easier to have my little store at the house. That way I can still

The sign for Shane Lucas's farm in Letcher County, Kentucky.

work in my garden and sell out of my store. Every year I keep adding something to it. Every single year it's getting bigger.

As we walked to the garden, the first thing Shane showed me was a thriving row of sweet peas.

SHANE LUCAS: I'm starting on my vegetables (pointing down the rows): there's tomatoes, eggplants, potatoes, cucumbers, cabbage. I've got twenty rows of peas through here now. I've picked a hundred pounds off of these, probably. Everybody knows me for my peas.

TOM HANSELL: People in Kentucky love their peas. Why is that?

SHANE LUCAS: I think it's a first-of-the-year thing. Come springtime, that's the first thing that comes out in the garden— everybody loves 'em. I'm picking cucumbers now, here's some of the little picklers . . .

Shane seemed genuinely happy in the garden but was still uncertain about his economic future.

SHANE LUCAS: Here in eastern Kentucky we don't have farmland like they do downstate, or in Tennessee. We have all hills. What flatland you got is taken. Nobody wants to turn it loose, or they building homes on it. We don't have acres of flat land to grow in. So now I'm trying to get into different things, growing blackberries, raspberries. I'm looking at growing mushrooms back in the mountains where there is a lot of shade. I didn't get to grow mushrooms this year. But next spring I hope to try to start the mushrooms. Maybe that could be a good business to supplement my vegetables and get something going that way.

After a quick snack of sweet peas, we walked over to a high-tunnel greenhouse.

SHANE LUCAS: I got about two hundred and fifty tomato plants in here. I hope within the next three to four weeks I'll be gathering tomatoes out of here. They're starting to fill out good. I've got some bigger ones in here somewhere . . . some of them are still blooming, too. Yes, they're loaded.

As I observed the tomatoes, I noticed that the large metal stakes he was using looked familiar.

TOM HANSELL: It looks like you've used some of your coal-mining heritage right here. Explain what these tomato stakes are made of.

SHANE LUCAS: These are roof bolts from a coal mine. This is what they bolt the top with to keep the rock from falling. They drill a hole in the roof, screw this in, and that's what they hold the top with.

TOM HANSELL: Now it looks like it's making a good tomato stake. Were those from a mine that has been shut down?

SHANE LUCAS: Yeah, they were going to bury them, cover them up. Got to salvage what you can to make it.

I'd like to have enough to do my whole garden, but I'm doing about 90 percent of it in roof bolts now. You don't ever have to worry about them rotting. I had been using wooden stakes, but I got these and it's the only way to go. They are a little bit heavier, but you can use them year after year.

TOM HANSELL: Many miners have a hard time wrapping their head around farming. I've heard some miners say that farmers are just another kind of tree hugger that are trying to put them out of a job. I'm wondering if you can tell me a little bit about your experience with that.

SHANE LUCAS: I'm a coal miner. A lot of the coal miners do relate farming to tree-hugging, a lot of them won't buy off of us because of that reason. They feel like they're supporting the tree huggers. I try to explain to them that it's not.

You know I'm still a coal miner. If a job popped up open tomorrow coal mining, I'd go right on back to coal mining. There's good money in it. But, coal mining is pretty well out, so I'm gonna try to do my best at farming, see what I can do with it.

Shane Lucas holds his granddaughter outside his shop.

REFLECTIONS

These are just a few of many innovative, hard-working people and organizations dedicated to community regeneration in the Appalachian
coalfields. It was a challenge for me to figure out which ones belonged
in the documentary, and continues to be a challenge to figure out
which stories belong in this book. During the production of *After Coal*,
I was continually heartened to learn about the broad range of groups
doing economic development work in coalfield communities. Already
hundreds, if not thousands, of people are hard at work reclaiming
abandoned strip mines to grow hardwood or industrial hemp, creating
local food economies, teaching laid-off miners computer coding skills,
and developing businesses for the region's growing technology sector.
Since completion of the documentary, I have learned of more promis-
ing efforts each day, more than I could fit into one book.

Still, there is much work to do, and serious challenges remain for
those working to build a sustainable economy in the coalfields.
History has shown that the familiar practice of economic development through industrial recruiting simply does not work in central
Appalachia. There is also the problem presented by the large percentage of land owned by absentee corporations. People need to find
solutions that work on a local scale, that match the specifics of their
local situation. Much of this work is unrecognized. I hope this book
and the documentary can offer at least a small amount of recognition to these heroes and help support their work. I also hope that
the stories in *After Coal* provide ideas and inspiration for others who
will carry on this important work of helping to find a way forward
after coal.

Tanya Turner and Elizabeth Sanders visit the DOVE Workshop in Wales.

Appalachian musicians Trevor McKenzie and Rebecca Branson Jones meet with Welsh musicians Frank Hennessy and Iolo Jones.

Richard Davies outside the Appalshop building in Whitesburg, Kentucky.

Nigel Davies and Chris King perform live on Tanya Turner's show on WMMT-FM.

CHAPTER 6

THE NEXT PHASE OF THE EXCHANGE

"APPALACHIA'S BRIGHT FUTURE" CONFERENCE AT THE HARLAN CONVENTION CENTER

HARLAN, KENTUCKY: APRIL 19, 2013

Eventually, the election of 2012 was over and Barack Obama was reelected for a second term as president. Although the Appalachian coalfields overwhelmingly supported his opponent (Mitt Romney won more than 80 percent of the vote in Harlan County, Kentucky), the toxic atmosphere of the campaign had faded, creating the opportunity for more dialogue (Politico 2012). In the wake of the election, I thought it was important to do my part to help ensure that coalfield residents did not feel left behind. I still hoped that conversations with Welsh residents who had survived life after coal might help heal divisions in the community and provide a way for people to recognize their common bonds and articulate their hopes for the future.

In April 2013, Kentuckians for the Commonwealth sponsored a conference titled "Appalachia's Bright Future" in Harlan, Kentucky. The goal of the gathering was to explore grassroots ideas about how to respond to the rapid decline in coal employment. Their goals aligned with my intentions for the After Coal project, and when they asked me to help organize a panel discussion on the opening night, I was more than happy to oblige. The concept of a "just transition" provided a central principle for the gathering. Kentuckians for the Commonwealth described the principle in these words: "We believe it's essential that the transition to the new economy is a just transition—one that celebrates our culture and invests in communities and workers who depend on the old economy . . . Our goal is to develop opportunities for our people, for eastern Kentucky, to thrive" (Commonwealth 2013).

The "Appalachia's Bright Future" conference was held at the newly renovated Harlan Convention Center in downtown Harlan. This three-day gathering featured information on the changing economy, lessons from other regions that have gone through transition, and examples of how entrepreneurs and organizations have worked to regenerate a diverse range of communities. Conference speakers

TIMELINE

	APPALACHIA			
	"Wales and Appalachia: Coal and After Coal Symposium" at Appalachian State University in Boone, North Carolina.		Terry Thomas visits Kentucky where he speaks at the Global Mountain Regions Conference sponsored by the University of Kentucky's Appalachian Center, and then participates in MicroFest at Southeast Kentucky Community and Technical College in Harlan County.	Hywel Francis and Mair Francis travel to Kentucky to speak at "Appalachia's Bright Future" conference in Harlan, Kentucky.
	OCTOBER 2010	**MAY 2012**	**OCTOBER 2012**	**APRIL 2013**
SOUTH WALES		Tom Hansell, Pat Beaver, and Suzanne Clouzeau travel to Wales to record the first round of interviews.		

included guests from places that have been through major economic upheaval, including Newfoundland fishing communities, logging towns in the Pacific Northwest, water protectors from the Navajo Nation, and tobacco farms in southwest Virginia.

The opening session of the conference was titled "After Coal: Lessons from Wales and Appalachia." Hywel Francis and Mair Francis traveled from Wales to speak at this session along with Helen Lewis, Pat Beaver, and myself. I was a bit nervous to present alongside such esteemed scholars, and the fact that more than two hundred people had filled the hall did not help. This was also the first time that I had screened clips from the *After Coal* film in a community setting. Many of the people I had interviewed would see themselves on screen for the first time. When the time came, I took a deep breath and started. Here are some highlights from the presentation.

THE WELSH AND APPALACHIAN COALFIELD EXCHANGE (2010–16)

Hywel Francis and Mair Francis travel to Kentucky to present at a public forum on coalfields public policy at the Appalshop Theater in Whitesburg, Kentucky.

Richard Davies travels to Kentucky to present at a public forum on arts, youth, and community regeneration at Southeast Kentucky Community and Technical College in Harlan, Kentucky.

JUNE 2013　　　**OCTOBER 2014**　　　**JUNE 2015**

Tom Hansell, Pat Beaver, and Suzanne Clouzeau travel to Wales where they present short video clips from the *After Coal* documentary at the DOVE Workshop and the South Wales Miners' Museum, and record more interviews.

Tom Hansell returns to Wales with Elizabeth Sanders and Tanya Turner from Kentuckians for the Commonwealth. The group presents at the DOVE Workshop, Theater Soar, and Swansea University.

TOM HANSELL: I'm honored to be invited to present here in Harlan County. I want to begin by clarifying the title of our presentation. We call it "After Coal," but I want to state that coal is not going away entirely. Coal is such an important part of this region's history, economy, and culture that we can't just flip a switch and turn it off. That said, it is also true that coal can no longer provide the large-scale employment as it did for most of the past century. Most everyone agrees that we need to diversify the region's economy. The question is: How to create an economy that works for most local residents?

Tonight we'll listen to the thoughts of people who have wrestled with that question both in Appalachia and in Wales. Let me introduce Helen Lewis, Hywel Francis, Mair Francis, and Pat Beaver.

THE WELSH AND APPALACHIAN COALFIELD EXCHANGE (2010–16)

	AUGUST 2015	MAY 2016	JUNE 2016
APPALACHIA	Swansea University and Appalachian State University enter into a formal exchange agreement, allowing students and faculty to work and study at each institution.		*After Coal* premieres at the Seedtime on the Cumberland Festival in Whitesburg, Kentucky. Welsh musicians Chris King and Nigel Davies perform at the festival as well as at a special event at the Kentucky Theater in Lexington, Kentucky.
SOUTH WALES	Swansea University and Appalachian State University enter into a formal exchange agreement, allowing students and faculty to work and study at each institution.	The *After Coal* documentary premieres at the Hay Festival in Wales. A post-screening discussion reunites everyone who was involved in the exchange between Appalachia and Wales between 1974 and 1976. After the screening and discussion, Appalachian musicians Trevor McKenzie and Rebecca Branson Jones perform with Welsh musicians Iolo Jones and Frank Hennessy.	

HELEN LEWIS: Looking at present-day Appalachia in a global context, I wrote: "In this new phase of capitalist expansion, we find that Appalachia and rural America become like third world economies and share their problems, high unemployment, lower wages, environmental degradation, community destruction, increasing poverty . . . Public schools and social security are in danger of being privatized. For some the government is an enemy to be destroyed."

Yet many people in Wales and Appalachia have deep connections to place, to their homeplaces, and want to envision a new future without abandoning the past.

After Helen's statement, we screened a short video clip featuring some of the younger people we interviewed for *After Coal*. The video juxtaposed Lauren Adams from Lynch, Kentucky, and Geraint Lewis from Seven Sisters, Wales, talking about their connections to their homeplaces and their hopes for the future.

HELEN LEWIS: We cannot hide from the fact that we are part of a global economy, but we can work to be cooperative, helpful, and not exploitive. We live on a fragile planet—we are all spinning around together and need to come together to save us all.

TOM HANSELL: We are interested in changing the conversation from the narrowly defined debate over jobs versus environment to the broader question of what is good for local communities. The exchange between Appalachia and Wales that Helen Lewis pioneered in the 1970s provides an opening that may move the conversation forward.

PAT BEAVER: Last May, we went to Wales to document the Welsh approach to economic transition, or what the Welsh call "community regeneration." Here's what we found:

- There is no magic bullet—former mining communities in the Welsh valleys are doing better than in the 1980s, but population has dropped and employment remains low.

- The first step they took was the greening of the valleys, establishing government programs to clean up mine waste piles (called "tips"), acid mine drainage, and other sources of pollution.

- Government needs to be part of the solution. In order for it to be part of the solution, people need to be actively involved in the process of government.

- Grassroots groups also need to be part of the solution. The longest-lasting initiatives are where people started their own program to meet their own needs.

- Perhaps most importantly, energy is still an intensely debated issue. In some areas coal is making a small resurgence—the latest figures have about a thousand miners

After Coal director Tom Hansell (standing) presents alongside project advisors Mair Francis and Hywel Francis and producer Pat Beaver at the DOVE Workshop in 2013.

working in Wales. But a critical difference in Wales is the existence of strict reclamation laws, so that communities know that within a certain number of years, with strip mines or opencast mines, the contours of the land and vegetation will be restored.

HYWEL FRANCIS: Meanwhile, energy corporations and the Welsh government have made a huge commitment to wind power—spinning turbines are commonplace above the former mining valleys. The government has partnered with private power companies to develop these wind farms and is now exploring large-scale tidal power.

On the surface this all sounds good. However, these former mining communities have well-founded concerns about outside corporations developing these energy sources and not leaving anything behind in the community. When we look at the proposal for an industrial-scale wind farm with eighty-seven new wind turbines to overlook the Afan Valley, we must ask ourselves: Who benefits? Is it the local communities or the stockholders of some offshore corporation?

TOM HANSELL: As we make the *After Coal* documentary, we are considering a series of questions: How do industrial and economic forces shape culture and community? What happens when fossil fuels run out? How does a community regenerate itself once the natural resources that created its main industry are gone? And, how can lessons from these areas speak to other resource-dependent regions throughout the globe?

The "Appalachia's Bright Future" conference renewed my belief in the power of civil dialogue. The conversations between Appalachian coalfield residents and guests from other communities who had lost their industry were inspiring, and a broad and diverse range of community leaders from central Appalachia shared their ideas. Kentuckians for the Commonwealth member Elizabeth Sanders described the importance of these discussions: "It's clear to most people that rapid economic transition is already underway in eastern Kentucky . . . The debate we need is bigger than jobs. It's about the kind of jobs we want and deserve. We need to think longer term and make economic decisions that are good for our land and our people" (Hardt 2013).

A crew consisting of Pat Beaver (producer), Suzanne Clouzeau (videographer), and myself (director) returned to Wales during the summer of 2013. Inspired by the spirit of exchange initiated at the "Appalachia's Bright Future" event, we gathered at the DOVE Workshop and the South Wales Miners' Museum to show film clips of Appalachian residents working to rebuild their coal-mining communities and to participate in community discussions with local residents working on regeneration in the Welsh valleys. The conversations were different than those in Kentucky, but there were some interesting overlaps. As I watched our Welsh partners respond to the short film clips we shared, I decided to continue the exchange with a series of public forums in eastern Kentucky.

COMMUNITY FORUMS

In the fall of 2014, I worked with producer Pat Beaver and community groups in the Appalachian coalfields to produce a series of three public forums that brought people from Wales to discuss strategies for community regeneration in Appalachia. This exchange was designed to provide a global context for the local challenges faced by Appalachian coalfield residents. The forums also served to reinforce ongoing efforts to increase economic and environmental justice in the Appalachian coalfields. The Chorus Foundation generously provided financial support for this work.

The Elkhorn City Riverwalk is part of the community's efforts to diversify its economy.

Our primary partners in the planning and organization of the series included the Elkhorn City Area Heritage Council, Kentuckians for the Commonwealth, Southeast Kentucky Community and Technical College, and Appalshop/WMMT-FM. These partners each helped identify a theme for the conversation they hosted, which included homegrown tourism, public policy to support regeneration, and the role of youth and the arts in building healthy communities.

Before each event, I edited a short video conversation starter to help focus discussion around the theme of the event. Next, WMMT-FM radio staff produced a radio report that provided supplemental information about each forum and helped promote the event. The live forums were recorded, then broadcast on WMMT-FM in order to reach a wide audience of coalfield residents. Here are some of the highlights from each forum.

HOMEGROWN TOURISM FORUM AT THE ELKHORN CITY PUBLIC LIBRARY

ELKHORN CITY, KENTUCKY: SEPTEMBER 18, 2014

A lot of water had flowed under the bridge that spans the Russell Fork River in downtown Elkhorn City (population one thousand) since my last visit. This community and its people have been close to my heart for a long time. I made my first documentary, *The Breaks of the Mountain*, in 1999, about local efforts to use the rivers and trails surrounding Elkhorn City to establish adventure tourism in the area. The Elkhorn City Area Heritage Council formed during the production of that documentary. Today, this group continues to be a driving force for celebrating the unique cultural and natural heritage of the area.

The world has changed a great deal in the fifteen years since I finished *The Breaks of the Mountain*. Coal employment was on the decline even during the Clinton administration, but over the past two years the bottom has dropped out of the market for central Appalachian coal, resulting in major job loss. The local economy is just as tough as it was in 1999; some might say it is tougher. Several storefronts in Elkhorn City still sit empty, and the venerable Rusty Fork Cafe on Patty Loveless Drive closed its doors during the past year.

But many residents in Elkhorn City have plans for a better future. One bright spot is the Pine Mountain State Scenic Trail, a 110-mile-long trail on the ridge that forms the Kentucky/Virginia state line. A trailhead opened in Elkhorn City a few years ago, and many residents see the combination of the trail, the nearby Breaks Interstate Park, and the white water of the Russell Fork River as a natural foundation for a new local economy.

I arrived in Elkhorn City with a little time to spare, so I walked to the end of Main Street in time to watch a train of empty coal cars roll through town. I took a few photos and then headed to the public library, where the forum was to be held. The Heritage Council had reserved the basement of the library for this event, and I was pleased that about twenty-five people had come out to participate in the conversation.

After a brief introduction, I started off by showing a short video conversation starter that highlighted two examples of local tourism efforts in Wales. The first was Call of the Wild, the private enterprise run by Geraint Lewis featured in chapter 4. The next was the Glyncorrwg Ponds and Mountain Bike Centre, a publicly owned outdoor recreation center built on a former mine site also discussed in chapter 4.

After watching the stories from Wales, local residents shared their visions for the future and identified obstacles that may prevent their visions from being realized. A series of great ideas quickly surfaced, some as simple as improving signage so that visitors can quickly locate attractions. Others offered long-term plans such as building a ropes course and a training center on riverside property. The main obstacle identified by residents of Elkhorn City is the lack of access to development capital. Local banks are often wary of funding start-up businesses with a high rate of failure, and local government is strapped for cash.

After the screening, Tim Belcher, a local attorney and leader of the Elkhorn City Area Heritage Council led the discussion.

> **TIM BELCHER**: Was the discussion in Wales like here [in Kentucky]? Some people think that if you promote ecotourism, then you're against mining. If so, how did they overcome that?

> **TOM HANSELL**: I think it was a little bit different in Wales, because their mining was shut down within a few short years.

View from the Towers Overlook at Breaks Interstate Park near Elkhorn City, Kentucky.

This resulted in a movement that unified not just the miners but the mining communities.

TIM BELCHER: So it kind of took a catastrophic event to bring everybody together.

TOM HANSELL: They were forced together during the strike.

TIM BELCHER: Whereas right now, we're kind of able to hang out and hope for the better.

TOM HANSELL: What do you think are the biggest obstacles you face in Elkhorn City?

AUDIENCE MEMBER (MALE): I think our biggest obstacle is somebody with a wad of money who wants to start a business. That hasn't happened.

AUDIENCE MEMBER (FEMALE): But a lot of these people in Wales didn't have a lot of money.

AUDIENCE MEMBER (MALE): But they found the money. And the programs available to them are probably different than what are available to us because we're different countries.

AUDIENCE MEMBER (FEMALE): Well, I have a question. We don't have one person with a wad of money. We don't have the government helping us like they did in Wales. How does a group of different people like us reach some of these people who can do these things for us? How do you get that started?

TOM HANSELL: That is the big question I am looking at with this project, and I don't have a real easy answer. But, I do think that groups like the Heritage Council are creating a network to support things to get you all started. I don't think there's a set formula for the next step. I think you all are going to need to figure it out, and the step for Elkhorn City is going to be different than the step for Pikeville or the step for Whitesburg.

TIM BELCHER: I think a lot of it is the level of public interest. Right now there are only a few people that get involved with trying to make these things work, it's not a community-wide effort. When you get five hundred or six hundred people in the community, your politicians are going to listen to what you have to say.

On my way home the next day, I stopped at the Breaks Interstate Park, just outside of Elkhorn City, Kentucky. It was early morning, and a cool mist shrouded the rock formation known as the Towers that rise dramatically out of the Russell Fork Gorge. Their natural beauty caused me to reflect on the discussion at the forum. I realized that the Elkhorn City Area Heritage Council ensures that unique natural assets are at the heart of local plans to build a future after coal. This place-based approach is a crucial element for creating sustainable coalfield communities.

However, access to capital for development is just as important and much harder to come by. Finding funds for sustainable development is not just a local challenge for Elkhorn City, but a challenge faced by residents in rural communities throughout the US and Great Britain. To create a better future, people in places like Elkhorn City, Kentucky, or Glyncorrwg, Wales, will need to be creative about identifying local assets and building partnerships to access the funds they need to regenerate their communities.

COALFIELD PUBLIC POLICY FORUM AT THE APPALSHOP THEATER

WHITESBURG, KENTUCKY: OCTOBER 7, 2014

The next forum featured Hywel Francis and Mair Francis and focused on the role of government in community regeneration. At this time, Hywel Francis was serving as a member of Parliament, representing the Aberavon constituency in south Wales. Scheduling a visit around the parliamentary schedule was complicated, but we got lucky and were able to time the forum to coincide with a celebration of *After Coal* advisor Helen Lewis's ninetieth birthday the following week.

The Appalshop Theater had taken on a festival-like atmosphere the night of the forum. The local chapter of Kentuckians for the Commonwealth had put out a great deal of publicity for the event.

The fact that a member of the British Parliament was visiting the small town of Whitesburg had attracted several members of the local press. At the reception before the forum started, TV cameras competed with the refreshment table for people's attention.

During the forum Mair and Hywel Francis were joined by panelists Evan Smith from the Appalachian Citizens' Law Center and Robin Gabbard from the Foundation for Appalachian Kentucky. The group watched short film clips from *After Coal* and discussed how government policy could help create healthy communities after coal mines close. The panelists began by explaining their approach to community development.

> **HYWEL FRANCIS**: It is easy to say that we are very similar communities. The old joke is that we are two nations divided by a common language. More seriously, there are very fundamental differences between the countries of US and Britain, which have shaped our respective mining communities. First of all is our National Health Service, state run, single payer, and the brainchild of a Welsh miner named Aneurin Bevan. That is a very big difference between the US and Britain and that needs to be stated in bold at the beginning. Secondly is the state and public ownership of the coal industry between 1947 and 1992. That had an impact on wages and conditions but fundamentally on the health of the miners and their families. Those are two very important distinctions and differences.

Those two major differences were created by the strength of organized labor, especially the strength of the miners' union. Despite the fact that the coal industry is gone, that sense of community, of solidarity, what church people call fellowship, is as strong and vibrant as ever. But as community members, we have to work to preserve our own sense of community.

MAIR FRANCIS: I was a teacher during the miners' strike and taught in the schools in mining communities. The women during the strike were helping to support their husbands and halfway through they looked at themselves and asked: "What are we going to get out of this strike?"

We were learning a lot and standing beside the miners, but we wanted to build our skills as well. The DOVE Workshop started off as a simple idea, where women could get together to develop a sense of where they could go to improve their skills or gain new skills. Both men and women attend courses, but the organization is run by women and decisions are made by women. I think that's quite important in a very patriarchal mining community where men have always seemed to make the decisions. New voices and new ideas are needed to regenerate these communities.

EVAN SMITH: Appalachian Citizens' Law Center is a nonprofit law firm in Whitesburg. We have three main areas where we work: helping miners and widows get federal black lung benefits, helping mine safety whistle blowers, and then we represent people that have environmental problems that are often related to mining or natural gas drilling or something in that area.

Our work is certainly related to coal, but we see our work changing. As part of that we are very interested in what the transition that this region is currently going through means, and what we should be doing to make it work better for the citizens of this area. It is a big question and there's not a lot of easy answers, but it is something we need to think about because it's going on around us.

I went to an Ivy League school and I was going to learn how to fix it and move back here and just fix what is wrong. What I learned is that is a big undertaking and there's no magic bullet that will fix it. One of the things we need to do is look to other places that have gone through this so that's one reason I'm so glad we can look to a place like Wales with all the parallels of this region.

ROBIN GABBARD: We refer to ourselves as the Community Foundation. We provide resources and technical assistance to communities and individuals all over eastern Kentucky. Community foundations typically have served as a portal for charitable investments—a place where people could invest their own funds for a charitable purpose. Those funds are then placed back into the community in the form of grants supporting a wide range of strategies.

The communities develop our strategies. We started with community conversations at firehouses, post offices, where people gather trying to figure out the things that we want to see grow and sustain over the next twenty, fifty, or one hundred years. From those conversations we developed some grant-writing strategies, and over the last five years we've been really working hard at developing a fund that would support those strategies.

Today, that work has spread across eastern Kentucky. We also see ourselves as a neutral convener for conversations bringing together the public, private, and individuals who want to see the development of a place where we would like to see our children, and other people's children, choose to live.

EVAN SMITH: For the kids that are graduating high school now, it is no longer an option to get a good salary from a job in the mines straight out of high school. That's actually the reality that most of the global economy has faced for a long time. There are not a lot of places in the world where you can make

$75,000 with a high school degree. Jobs in mines are hard, I'm not saying that's an easy life, but it's been a way that people have been able to provide for their families without investing in as much education as a lot of other people in the national and global economy have had to.

Looking forward, we're going to have to teach our kids that you are going to have to invest in more education. That investment means not just attending school but also figuring out how to be engaged while you're there. If you're planning to start a small business you need creativity, skills, and a different mindset than *My dad worked in the mines, so he is gonna get me a job there*. That's not reality anymore, so we have to have the education and skills to figure out what makes sense in reality.

HYWEL FRANCIS: You cannot have a successful, prosperous business community unless you have skills. There is no doubt

After Coal project advisors Mair Francis and Hywel Francis speak with Robin Gabbard and Evan Smith at a forum on coalfield public policy at the Appalshop Theater.

about that. You can't start with a culture of miners who simply demand that factories should be arriving here on our doorstep to replace the mines. That's not going to be the case. We need to ask the question: How do you create a new economy that is in fact different, with lower levels of populations, smaller populations?

When we opened up the forum to questions, retired coal miner and mine inspector Stanley Sturgill from Harlan County stood up. I met Stanley when he was working with the local chapter of Kentuckians for the Commonwealth to protect the drinking water of Benham, Kentucky, from mountaintop-removal coal mining. His question highlighted his concerns about mine reclamation:

One of our biggest hindrances here in our area is the division between our people. You have so many people that say if I'm against strip mining or mountaintop removal then I'm branded, I'm a tree hugger, I'm called different names. There's just a division that's gonna have to be settled out before we can make progress.

I'd like to ask you, we have a lot of these mining industries that will skirt around anyway they can to try and get out of refurbishing these mountains. Whatever they've done to destroy them they don't want to do anything to correct what they have done. You mentioned your reclamation department required that you put a mountain back better than what it was; I'd like to know how to do that. God built these mountains, but what these coal companies leave us with is far from what God put there. I know we can't put them back like God did, but I would also like to know what you do require, what your government requires of them, and how you get it done?

The panelists replied:

HYWEL FRANCIS: We don't have the scale of mountain removal that you have. We don't remove mountains in that respect, certainly not in the same magnitude. These are much smaller developments, really just small excavations, compared with yours. I'm surprising myself, saying this now, but the land that I grew up on, the mountains that I used to play on, were mined. Very sparse vegetation was left, it was not a particularly beautiful landscape, but now that whole area has been vastly improved with much stronger grasses and a much wider range of trees, leisure facilities, mountain bike trails, and trails for walking. The whole landscape has been transformed and it's very, very green.

But to do this requires government intervention, which large parts of the American society don't like. A high level of authority is required to protect the environment. We learned a great deal from you with your national parks and what the Roosevelts did and that's admired all over the world. The park system is a legacy people should be proud of, because it is an example of the government working on behalf of the people for prosperity.

We shouldn't see the land as ours, it's for the future isn't it? It's an absolute disgrace that mountains are taken away in that way and the wealth that is removed. It seems common sense to me that they should have a responsibility to replace those mountains and put back, given that they've been lost to the community for that period, what is put back should be put back in a better sense and it should be accessible as well.

HELEN LEWIS: One of the things I think we have been misled on is the idea that if we will just get rid of the regulations then the coal industry will rise up again and we will have jobs. I think we just have to recognize that coal is gone and we can have new industries. First, we need to look around and see what we already have that we can use.

What are our resources and who owns the land? Although the coal industry may be closing, they still own a lot of that land. I talked to a winery over in Wise and he said they can't get any more land because the company will only lease it for one year. You can't plant a vineyard in a year. I think we have to find a way to get our resources back.

I spent a lot of time working in Wise County, in the coalfields. When I first came there in 1955 there were one hundred orchards, great apple cider and apples. There's one little orchard there now. Why can't we go back and get our land and do some thinking about new industry and find some ways of getting funding as reparation from all of the destruction that has been done?

EVAN SMITH: I agree with so many things that Helen just mentioned. It would require an act of Congress to somehow get money; I don't think that's something you can just do through the court system. But, there are sources of funds that are out there that you could use to restore the area and infrastructure.

For example, there is a federal Abandoned Mine Land Fund. There's a lot of politics involved with all of this, but there is $2.5 million in this fund that is meant to be used for the legacy costs that come from mining. We need to find a way to get some of those funds, put them to use, and be smart about how we use them. We could extend water lines, turn old mines into something you could grow apple trees on. When we do this, we also want to make sure we're using local contractors, that way we can make each dollar circulate as many times as possible and generate future dollars. I think we really need to think about how can we have jobs that are going to be here for generations to come.

This is not a regional, or even a local problem. We are a part of the United States of America, our residents fought for this country, and I think that the national government has an obligation to help us get our economy back on its feet. When military bases close, the federal government helps people get jobs. When the timber industry was changing in the Northwest, the federal government got involved in helping those economies transition.

We're so used to thinking in this area that Washington, D.C., is the problem and that we are being besieged by bureaucrats, but I think that, although Washington, D.C., represents the problem, we also have to think of them as a solution. The Abandoned Mine Land Fund is run by Washington and we need to focus on how our local elected officials that are in Washington, D.C., can effectively represent our area. I think that we have to think about how we can form alliances with people in other regions if we're really going to be successful in D.C., and I think we have to be if we're going to be successful here.

The Welsh community also provides hope that we should imagine a future that is not filled with despair. As long as we can plant gardens every year, as long as we can have grandparents playing with grand-babies, there's going to be hope.

—Evan Smith, attorney at the Appalachian Citizens'
Law Center

ARTS AND YOUTH FORUM, SOUTHEAST KENTUCKY COMMUNITY AND TECHNICAL COLLEGE

HARLAN, KENTUCKY: OCTOBER 28, 2014

The final forum of 2014 focused on the role of arts and youth programs in community development. During my visits to Wales, I met Richard Davies, who directs the youth media program at Merthyr Tydfil College (a two-year college in a former mining town). He was very interested in the Higher Ground project, so I invited him to present at the forum in Harlan, Kentucky. Davies presented work created by his students and shared lessons learned from converting the old town hall in Merthyr Tydfil into an arts center for the college. Students from Higher Ground showcased their work and talked with Richard Davies about the role youth can play in revitalizing former coal communities. Devyn Creech, one of Higher Ground's youth leaders, started the conversation:

DEVYN CREECH: I grew up thinking that coal was forever. That was my ideology as a young girl. Both of my grandparents worked in coal mining and my dad was a coal miner for a while, to me it seemed so stable. After high school, I couldn't wait to get away from here, so I moved to Morehead, Kentucky, for college. It was just three hours away [from here], but I was absolutely miserable.

When I came back home, I was asked to join Higher Ground and it got me really, really invested in this place and made me realize how much potential we have to offer as a community. For me, this project is so much about a love of place and helping the people that I love. I just think that with the arts we have a means to open the floor to people and let them understand that we can make a space to talk about these economic issues and community issues. We don't have to always argue, we can just kind of open up for each other.

RICHARD DAVIES: I am so pleased to be here. I never expected to visit Appalachia, let alone America, which is kind of awesome. I'll start telling you a little bit about south Wales, and about Merthyr Tydfil in particular. There are such parallels between south Wales and here over the last thirty years that you may be seeing in your community.

Merthyr Tydfil has had three different periods of deindustrialization, different changes in the industry. Each time the wave has passed over but, of course, Merthyr Tydfil is still there, south Wales is still there, and I think that's really important to remember. What you need to know is how has the place survived.

I'm from Merthyr Tydfil, I was born in Merthyr Tydfil, and one side of my family has lived in Merthyr Tydfil since before

Devyn Creech introduces the "Arts and Youth" forum at Southeast Kentucky Community and Technical College.

it industrialized, so they came from a farm. I am involved in filmmaking; I produce programs reflecting on the nature of Merthyr Tydfil, and the kind of society that [has] developed there. During the miners' strike I became involved in recording people's memories and struggles, not exclusively about mining but about the kind of communities that had been shaped by the experience of the coalfield.

In 1985, when the strike ended, it was clear that mining wasn't going to continue. It became clear that the government was determined to close collieries; some continued until 1991 and then that was it. What happened after was that people almost went into a period of mourning, I suppose. I think you could put it that strongly. Coal had become central to people's identities and sense of themselves. So the idea that the village that had always been there because of the mine, the idea that coal could go away really, really shocked people.

Richard Davies presents at the "Arts and Youth" forum at Southeast Kentucky Community and Technical College.

So there had to be a response to that, and my response was to try and find out what it meant that Merthyr Tydfil could survive the previous deindustrializations. What factors meant that Merthyr Tydfil could survive as a town? One of the things that I identified was music. People sing, not only in organized groups, but also [there's a] tradition in south Wales of something called an informal hymn singing where a chapel will go to another chapel and they will spend the evening singing hymns. I love going and recording them. I have a load of tapes of chapels holding these informal singing sessions. There's a very strong tradition in Merthyr Tydfil of folks singing ballads in the Welsh language. They probably haven't traveled very widely, but they are very interesting and the issues they're dealing with are the issues that exist now. Music can allow us to see how people responded to similar kinds of crisis in the past.

The next thing becomes interesting because it becomes to be about the future. If you know about local music, arts, and culture, how can you make them live, work, do things for you now?

So that led to my involvement with young people. It's amazing to see the kind of stuff that is going on with Higher Ground here, because some of the concerns that are expressed in the Higher Ground project are expressed by my students in Merthyr Tydfil.

I think you've got to follow what young people are saying in these communities. Sometimes older people don't necessarily want to hear that, but you have to keep on saying it. With the declining coal, the declining industries as well, who does that affect? It affects young people and their futures in these places.

—Richard Davies, director of the youth media program at
Merthyr Tydfil College

I decided what I had to do was to begin working with young people. I believe that you've got to get people's qualifications up and give them a chance to develop themselves so I started teaching at Merthyr Tydfil College, which is very similar to this community college. It's a two-year college, we start with people who are fourteen to fifteen years old who are not in school and we work with them and produce films. It is incredibly like this, when I walked in, this sense and atmosphere of being at home, I suppose. I recognize the kind of posters on the walls, the titles of the rooms, the description of the courses, so again that made me think we should be talking to each other. We should be exchanging ideas and thoughts about all of this because you can help me in what I do and I think my students and the people I work with could help you.

Richard shared two short videos produced by youth in Merthyr Tydfil. One documented the transformation of the old city hall into a night club, which was then abandoned, only to be renovated as the home of the media program for the community college. The other profiled a local man of Italian descent who was also an Elvis impersonator.

The films went over well. After the screening, Devyn Creech and Robert Gipe from Higher Ground asked questions that deepened the conversation about how youth and arts programs can support economic regeneration.

DEVYN CREECH: How many people live in Merthyr?

RICHARD DAVIES: About thirty thousand in the town. You've got the town of Merthyr but you've also got the Merthyr borough, which includes five or six other communities, former coal-mining communities.

DEVYN CREECH: What is the industry there now?

RICHARD DAVIES: Mainly service industry, the largest employers are the college and the hospital. Recently a car exhaust manufacturer just opened up; their factory employs two hundred and fifty people, which would be about the largest employer. When coal went, we lost a lot of high-paying jobs. Coal paid better than anything else, and the loss of the income was destructive.

ROBERT GIPE: Sounds like the parallel between our communities—the only place you can get a good job is in health care or in the educational facilities, that's all that's left.

Let me ask you something: The coal industry collapsed in 1980-something? So we should just buckle down for thirty more years, at least?

RICHARD DAVIES: I think it's a process. Some things get better sometimes, and sometimes things just don't get better. It's not a constant onwards and upwards to rise and recovery. It's bits and pieces that start getting better. There's tremendously inspiring work being done by people in my town, and I think one of the things that appears to really influence whether their projects are successful is how closely they're connected with the communities they claim to serve. Projects that are started by the people that live in them appear to be much more successful than most things that come in from the outside.

ROBERT GIPE: I think the real challenge is to figure out some organic economy for this work. Tom and I have been doing this work for about twenty-five years in these communities and it feels more fragile to me than it ever has.

On May 26, 2016, all of the original participants of the Welsh and Appalachian coalfield exchange during the 1970s gathered at the Hay Festival at Hay-on-Wye in Wales to participate in a panel discussion that followed a screening of the *After Coal* documentary. This was the first time in forty-two years that the entire group had been together. Dai Smith, former chair of the Arts Council of Wales, facilitated the conversation.

DAI SMITH: I'm going to begin, if I may with John Gaventa. Why did you come into south Wales in 1974?

Above:
After Coal panel discussion at the 2016 Hay Festival at Hay-on-Wye, Wales. From left to right: John Gaventa, Mair Francis, Hywel Francis, Dai Smith, Helen Lewis, Richard Greatrex.

JOHN GAVENTA: As a student I had a scholarship to come study in Oxford after I had been working in a mining community in Tennessee. Someone gave me the name of a filmmaker, a very well-known filmmaker named George Stoney in New York, and he gave me this contraption called a video machine. Back then it was the old reel-to-reel recorder, and it was quite heavy and bulky. Stoney asked: "Could you experiment and see if you could find any lessons to share back in the mining valleys here in Appalachia?" But I knew nothing about filmmaking and I arrived here in Wales in the middle of the national miners' strike of 1974.

Somebody had given me Hywel's phone number, so I called Hywel and asked: "Do you mind if I come over and talk to some miners?" He said: "Sure, come over." I remember Hywel set up an interview on a Friday night, and the miners were having a great time,

but they kindly took some time off to be interviewed by this young, very naive student from America. And I remember my first question was: "Tell me what led up to this strike?" And of course I knew nothing about history, especially the importance of [the] history in south Wales. When I asked that question, all the younger miners put their heads on the table and one older miner said: "Well, son, it all began in 1926 . . . "

And about an hour later, my battery was dead, I had no more equipment, half the miners were asleep, and we were only up to 1945! So that's the way it started, just that very simple way. From that the whole project grew, and the exchanges continued for the last forty-two years.

DAI SMITH (to Hywel Francis): In the mid-1970s, did you think that the labor victories meant a beginning of a better life in the Valleys?

HYWEL FRANCIS: Reflecting back, I think that the 1970s were an unusual period, both in the US coalfields and in the British coalfields. It was a kind of a respite from the long decline that began in the late 1950s and continued through to the 1960s. Suddenly, you had in both countries the tripling of oil prices. As a result, coal was very much in demand and the miners' unions therefore were politically much more significant. Labour had sufficient economic power to actually achieve a political change. That was a consequence of the world change in the price of coal.

Similar struggles were going on in the US. Suddenly, the miners were much more center stage due to President Carter's drive to support coal. Of course the big argument in America was not so much the mining of coal, but where the mining was going to take place. Our understanding was that it was going to move from the eastern unionized coalfields to strip mines in the Midwest. These strip mines were largely nonunion.

The one thing that combined and joined us all together was that we were all engaged in one way or another in adult education, and quite radical education. We each saw a link between radical education and social change. That's why we were so pleased to start working with John throughout that period, and it resulted in exchanges which last today. These exchanges are now in fact, much stronger than ever, despite the fact that the coalfields have all but gone.

DAI SMITH: Well, isn't that the point? However, despite those moments of hope, very shortly in Wales and Appalachia we were, in some sense, managing decline. Helen, please tell us how you got involved.

HELEN LEWIS: I moved to the coalfields of Appalachia in 1955.

It was a time that they brought in the mechanized miners, and people were leaving in leaps and bounds. I was just really interested in understanding the dynamics of this place. I had grown up in the mountains but not in the coalfields, so I was really intrigued by the coalfield culture and all the people who were having to leave because of the mechanization of the mines, so I decided to research coal mining, and what it did to communities. So that's why I got started.

DAI SMITH: Then when you came to south Wales, did you feel that the comparative thing was working straightaway? You talk about a coalfield culture; did you conclude that there was one?

HELEN LEWIS: Well, I think it probably is similar to other dangerous jobs where people have to depend on each other to save their lives. The comradery that binds together the coalfield communities and

families in Appalachia was really impressive, and I really wanted to see if that happened in other coalfields. That is when I started looking for a possibility of visiting other places.

DAI SMITH: What did you think was the biggest difference you found between what the south Wales miners were going through and miners in Appalachia? What was the biggest difference, really?

HELEN LEWIS: The biggest difference was that in Appalachia, the coal companies owned our communities. In Wales, you at least had control of your own houses. That made a big difference. The nationalized coal industry in the UK was very different than the complete private ownership in the US. So I wanted to bring miners from Appalachia to Wales. Get some exchange going. John and I thought that if we could get the miners talking to each other in both countries, they

*After Coal panel
at the 2016 Hay Festival.*

could understand what the relationship was between who owns the mines and who owns the community.

DAI SMITAH: At a rank-and-file level with the miners, did the exchange work? Beyond you researchers and educators, did the miners actually bond?

HELEN LEWIS: Yes, they really got interested in talking to each other and understanding the differences. When I came over here to Wales, I brought some miners from West Virginia over, and they visited all the mines and started to develop relationships. Then when I went back to Appalachia, John and I had miners from Wales come over, and we took them on a tour, and talked to miners and their own communities. In addition to bonding, people really got to talking about the differences between the two systems.

DAI SMITH: Richard, what about imagery—how the landscape has been changed

by coal? Is this what interested you as a filmmaker?

RICHARD GREATREX: It's difficult to say. Because back when, in 1974, what did I know? Very little. I mean, the reason that I got involved was that I knew where south Wales was. I'm originally from Swansea, so I could show John where it was down the motorway, so that helped. But what did I know about the mining communities? What did I know about cameras at that time? I knew very little.

However, I had been trained as an electrician. So what I did know was how to make the equipment work. I could make the reels of videotape go round and round when they didn't want to go round and round. This happened all of the time, because we were taking the camera and recorder places they were never meant to be— like in a coal mine. So my input was that I could keep the show on the road.

I've learned about imagery

in the thirty years since. But what was important for me, and this is a very personal thing, is that this project introduced me to filmmaking and allowed me to understand what a cinematographer could do, and that has been my career ever since. So, it's down to these guys that I actually found a job that I really liked, rather than be an electrician.

DAI SMITH: When you were making the machinery to go round and round, at some point, didn't it occur to you at some subliminal level that maybe you were making something happen that no one else was doing?

RICHARD GREATREX: Absolutely, and there was a lot to be learned. Interestingly enough, a lot of what I personally learned was not about the image making, but about how you get to the images—especially how you get the repartee with the people in front of the camera.

It wasn't just a matter of putting a camera in front of them, you have to create the right kind of situations to get those images. First of all you have to put people in a comfort zone. That's why we took people to places they are familiar with, or talked in their front rooms. But another thing is you have to challenge people. You couldn't let people give you superficial answers. You have to try and go deeper, and get under their skin a little bit. That was the major thing

After Coal project advisor John Gaventa speaking to Mair Francis at the 2016 Hay Festival.

I learned from my work with John and Helen.

HYWEL FRANCIS: Part of what Helen and John did was to make us more aware of what was happening in the United States. I never thought I would get to visit the United States because I had been a member of the Communist Party. But Helen and John made it happen, and I went to the US in 1979 with six miners. In our odd, patronizing way we thought that they were way behind us, but what we actually saw was the future.

We saw terrible conditions, terrible working conditions: unbridled exploitation of miners, nonunion coalfields, no local democracy. And we thought: "Well, we've seen the past. They're still struggling with what our fathers and mothers defeated." But in fact, within five short years, we were actually facing exactly the same circumstances through an imported coal board director in the form of Ian MacGregor, who treated us the same way the miners were treated in the US.

JOHN GAVENTA: I think one of the things that impressed us most when we first came was the difference in politics here, especially the difference in power the trade union movement had built through collective organizing. The trade union movement was also partnered with a strong adult education movement. It was such a contrast for me between miners getting shot at on the picket lines in Harlan County, then arriving in Wales and watching tens of thousands of miners marching in the streets. Miners had the power and could help bring down a government at that time. So it was a huge contrast. These cultural differences fascinated me at the time and I am glad that the exchange has continued.

Appalachian musicians Trevor McKenzie (far left) and Rebecca Branson Jones (far right) rehearse with Welsh musicians Iolo Jones (center left) and Frank Hennessy (center right) in the studios of BBC Radio Wales.

RICHARD DAVIES: I think that's very true. Anything that I do is like a drop in the ocean. But I still have to do it because my grandfather did it, my father did it, everyone has fought for the community, for their place, and if I don't do that now, then I've got daughters . . . what are they going to grow up into?

I think we're asking young people to hang in there and I think they'll be seen by future generations as the most heroic people that this place has ever seen, because them staying in this place is making a place for them. I admire them and I'm here because they're here at this point.

—Robert Gipe

A MINER'S LIFE *(lyrics to a traditional folk song)*

A miner's life is like a sailor
On board a ship to cross the waves
Every day his life's in danger
Many ventures being brave

Watch the rocks, they're falling daily
Careless miners always fail
Keep your hand upon your wages
And your eye upon the scale

Union miners, stand together
Do not heed the owner's tale
Keep your hand upon your wages
And your eye upon the scale

You've been docked and docked again, boys
You've been loading three for one
What's the use in all your working
When your mining days are done?

Worn out shoes and worn miners
Blackened lungs and faces pale

Oh, keep your hand upon your wages
And your eye upon the scale

Union miners, stand together
Do not heed the owner's tale
Keep your hand upon your wages
And your eye upon the scale

In conclusion, bear in memory
Keep this password in your mind
Worker's strength cannot be broken
When unions be combined

Stand up tall and stand together
Victory for you prevail
Oh, keep your hands upon your wages
And your eye upon the scale

Union miners, stand together
Do not heed the owner's tale
Keep your hand upon your wages
And your eye upon the scale

On our way to the airport, Richard and I agreed to talk more about the next phase of the exchange when I returned to Wales the following summer. In the meantime, I made arrangements to introduce two members of Kentuckians for the Commonwealth, Elizabeth Sanders and Tanya Turner, to my Welsh partners during the trip. I returned to Wales in 2015 and screened a fine cut of the documentary to our partners at the DOVE Workshop. Elizabeth and Tanya helped facilitate the post-screening conversation.

Richard Davies had also arranged for a work-in-progress screening of *After Coal* at a historic chapel in Merthyr Tydfil the next day. After the screening, the four of us walked to the New Crown Pub to talk about ideas to continue the exchange. Over a few pints we cooked up a plan to present music from Wales and Appalachia in conjunction with screenings of the *After Coal* documentary in both regions.

Musicians from Appalachia and Wales harmonize at the 2016 Hay Festival. From left to right: *Iolo Jones, Frank Hennessy, Trevor McKenzie, and Rebecca Branson Jones.*

2016 MUSICAL EXCHANGE

The first event in the musical exchange occurred on May 24, 2016, when the *After Coal* documentary premiered at the Hay Festival in Hay-on-Wye, Wales.

For the UK premiere of the documentary, I traveled to Wales with an entourage that included producer Patricia Beaver, advisor Helen Lewis, and two young Appalachian musicians: Trevor McKenzie and Rebecca Branson Jones. Trevor and Rebecca were scheduled to perform with Welsh musicians Frank Hennessy and Iolo Jones. We had worked with Andy Fryers at the Hay Festival to set up three related events: a film screening led off the series, followed by a panel discussion featuring all the people involved in the 1970s exchange between Appalachia and Wales. The musical exchange, billed as "Songs from Both Sides of the Atlantic Coal Seam," would conclude the series. Swansea University helped sponsor these events.

The four musicians met for the first time at the BBC Wales studios in Cardiff a few days before the festival. They had shared song ideas via email, but this was the first time they had the opportunity to play together. They started by kicking around a few songs they knew in common, introducing fast fiddle tunes like "Sally Goodin" to her Celtic cousins. Then they worked on vocal harmonies for some of Frank Hennessy's original folk songs, including "Farewell to the Rhondda" and "Tiger Bay."

Via email the group had agreed to develop a medley based on the Welsh hymn "Calon Lân." Frank told us that this hymn had been brought across the Atlantic more than a century ago, where it evolved into "Life's Mountain Railway"—a bluegrass gospel standard also known as "Life's Railway to Heaven." Eventually, the tune returned to Wales, and miners' choirs changed the words, creating "Miner's Lifeguard" (also known as "A Miner's Life"), now a standard in the repertoire of

Welsh male voice choirs. The musicians' goal was to create a musical arrangement that traced the hymn across the Atlantic and back again.

Under Frank Hennessy's astute musical direction, the quartet reversed the song's history and created a medley that started with the entire group playing "Miner's Lifeguard," then reverted into an a cappella chorus of "Calon Lân" before Rebecca's banjo brought in "Life's Railway to Heaven." The chemistry between the musicians created an electric atmosphere in the studio, and the entire process of arranging the medley took less than thirty minutes. I felt fortunate that I was able to watch the magic happen.

The group gathered at the Hay Festival early on the day of the event. As I went through the tech check in the tent reserved for our film screening, the musicians were guided to another tent, where their medley was broadcast live on BBC Radio 4's *Front Row* program. When I regrouped with the musicians after their show, the festival organizers informed me that the musical performance had sold out and that the musicians would be moved to a larger tent. Apparently, the musical exchange was more popular than the film!

After the screening and discussion, we moved to the big tent for the music. First Frank and Iolo warmed up the crowd, then Rebecca and Trevor took the stage. Finally all four musicians came together and swapped tunes from Welsh and Appalachian traditions. The climax of the evening came when they performed the medley; the entire audience sang along to "Calon Lân." As the last note of the hymn faded, you could have heard a pin drop in the big tent. After a single beat of silence, Rebecca kicked her banjo into the chorus of "Life's Mountain Railway" and the quartet joined in in perfect harmony. They got a standing ovation, and the crowd would not let them leave until the festival organizers shut the show down.

Welsh musicians Chris King (center) and Nigel Davies (right) swap songs with Appalachian musicians Michael and Carrie Kline (left) at Appalshop's Seedtime on the Cumberland Festival.

BACK IN KENTUCKY

The following week, we brought the musical exchange back to Appalachia for Appalshop's annual Seedtime on the Cumberland Festival 2016.

I picked up Richard Davies, along with Welsh musicians Chris King and Nigel Davies (no relation to Richard) at the Lexington, Kentucky, airport well after midnight. But the next day everyone was up early and ready to go. After a quick stop at the Doo Wop Shop to purchase new guitar strings, we hit the road. We arrived in Whitesburg that evening in time to experience the Feedtime on the Cumberland, a local food event at the Letcher County Farmer's Market before the Seedtime on the Cumberland Festival.

The next morning Chris King and Nigel Davies were guests on Tanya Turner's radio show on WMMT-FM and shared a few of their original songs. That evening, we screened *After Coal* and hosted a public discussion with Richard Davies about the path Wales took after coal. Then we all went out to the big tent to celebrate with a local honky-tonk band called the Giant Rooster Sideshow. The band sounded so good, we kept moving closer until we were almost onstage with the group. Then, in a moment of wild abandon and international solidarity, Chris ended up briefly playing the lead guitar with his tongue!

As we were leaving the hotel the following morning, we met West Virginia musicians Michael and Carrie Kline in the hotel parking lot. They invited Chris and Nigel to join them in an informal jam session under the apple trees that line Appalshop's grounds. As I watched the quartet swap songs, I marveled at the universal language of music.

We all drove back to Lexington the next day to set up for an event at the Kentucky Theater. From talking to the theater's technical director, I understood that we had to provide our own sound system for the musicians, but when we arrived at the theater, we also saw that we needed to provide our own lights. Fortunately, Richard remembered seeing lighting equipment at the Doo Wop Shop when we bought guitar strings the first day of the trip. We returned to the shop, and after convincing the staff that they should rent to a group of Welsh musicians, we acquired the lights. The event was a great success—more than three hundred people attended and it was a fitting conclusion to the musical exchange.

The following day, I dropped the trio off at the airport, drove the seven hours home, and slept for a week. I was pleased to have helped organize a series of events that underscored common musical traditions and the strength of music to bring people together across cultures.

Landscapes of the central Appalachian coalfields (top) and the Welsh coalfields (bottom).

Landscapes of the central Appalachian coalfields (top) and the Welsh coalfields (bottom).

CHINA

BY PAT BEAVER

In May 2013, I had the opportunity to travel to Shenyang, China, to talk about the After Coal project. Shenyang is in Liaoning Province, in the northeast, and is a major coal-mining area of China. My family and I lived in China in 1983–84 and again in 1990–91. This was my first return to experience the radical changes in China over the past decades.

In 1983 China was dismantling the communes and economic reform was a goal. Many of Shenyang's three million residents cooked and kept warm with small stoves fueled by pressed coal briquettes, not unlike previous decades in Wales and in Appalachia. Early mornings saw millions of bicyclists making their way to work, their faces shielded with cotton masks against the coal smoke that hung heavily over the city. Automobiles were company-owned and rare, although that year saw the first private car purchased in China. Motorcycles and hit-and-miss truck engines added a level of cacophony to the compelling sounds of bicycle bells. I was teaching at Northeast University of Technology; my students had been selected for training as engineers for the mining industry.

Returning to China last month I was astonished by Shenyang's modernity, as high-rise apartment buildings, skyscrapers, billboards advertising the latest fashions, and glittering shopping malls dominated the landscape. Yet modernity mingled with shadows of the recent past, with dimly lit night markets selling modern fashions alongside live fish, fruits and vegetables, and backstreet vendors selling dumplings steamed over gas-fired street burners.

Like Appalachia and Wales, Shenyang is powered by electricity generated from coal-fired plants. Shenyang's streets are clogged with privately owned cars and motorcycles competing for lanes with trucks, buses, and a small number of brave bicyclists. High-speed trains, four-lane highways, and air transportation connect city to city. China's expanding wealth is evident along the streets and in the classrooms. While the early morning pall of coal smoke has disappeared, the pollution is visible in the haze that caps the city and the region.

In Shenyang I talked about the After Coal project to faculty and students studying engineering and English. I concluded by mentioning the article I had just read in the *New York Times* declaring that the level of carbon dioxide in the earth's atmosphere had reached 400 parts per million, the highest percentage since we have become human. As I talked about the declining coal production in Appalachia and Wales, I asked: "What happens to communities when coal mines close? How can communities rebuild, or regenerate, after they lose their major industry? Is there a sustainable future for former coal communities?"

The story of Appalachia and of Wales is relevant to Shenyang and to Liaoning Province because of coal mining in Fushun, the location of the largest open-pit mine in China, and Fuxi, where the mines have closed. This has resulted in a dramatic loss of jobs, loss of businesses as people move away to seek work, and an uncertain future.

I also learned that engineering students in this region are looking at clean coal technologies, not alternatives to coal. China is making choices about its energy sources and is pursuing development on a massive scale. Our futures, in Appalachia and in Wales, are linked to those decisions.

CHAPTER 7

CONCLUSIONS

COAL AND AFTER COAL

People sometimes think that we are living in a period of after coal, that coal mining has come and gone. I would challenge that view. The electricity coming to power these lights is coming from the Aberthaw Power Station, which is coal fired. That coal is coming in the main from Russia, from Columbia, from parts of the States, from China. We don't supply enough coal to meet the demand in Wales or indeed the UK. So we import a huge amount of coal every year.

—Dean Cawsey in the *After Coal* documentary

As I look back on the process of making *After Coal*, I realize that many loose ends remain untied. Although the goal of the documentary was not to offer answers, but rather inspire conversation, several complicated (even contradictory) points linger. For example, the project and this book are titled *After Coal*, yet coal has not completely departed from Appalachia and Wales.

Many people in coalfield communities rightfully point out that coal is still an important part of electricity production. However, coal's economic contribution to the communities it once sustained has all but disappeared. Both the US and the UK have increased their imports of cheaper coal from other countries at a time when demand for coal-fired electricity is declining. For example, in the US, coal's portion of the electric power pie has dropped from more than half in 2000 to less than one third in 2016 (USEIA 2015). In the UK, coal-fired power generation dropped sixty-six percent in 2016, and the government plans to phase out all coal-fired power plants by 2025 (Vaughan 2016).

The playing field in energy producing communities is changing. Large industrial unions such as the United Mineworkers of America and the National Union of Mineworkers are no longer able to influence national policy. Recently, teachers and public employees have had some success developing new labor organizing strategies. However, the question of how to develop thriving working-class communities remains unresolved. As Terry Thomas reminds us at the end of the *After Coal* film: "We have to find new ways to fight for our communities."

The question then becomes: How can community members participate in building an economy from the bottom up instead of being stuck in the top-down model favored by globalism? Expanding region-to-region exchanges such as the ones documented by *After Coal* may be part of the answer.

GLOBALISM VERSUS INTERNATIONALISM

When people talk about the antithesis, the opposite of globalization, the word they use is internationalization. *Not international corporations exploiting the world, but people creating solidarity locally and internationally.*

—Hywel Francis (Francis 2010)

One of the lessons that I learned through the production of *After Coal* is that coalfield communities have a surprising supply of local resources at their disposal. Looking inward and developing transparent, democratic mechanisms to control local resources is a first step communities can take toward surviving the rapid changes of a globalized economy.

However, communities need to be careful to temper this local loyalty with a global perspective. A narrow focus on your local community ignores the fact that we're part of a larger world. We can't put our collective heads in the sand and pretend that we are not connected to a global economy. One of the reasons both Appalachia and Wales lost jobs so quickly was that coal markets were responding to an influx of cheap coal from overseas, where policies to protect workers and the environment were basically nonexistent. This lack of protections made it cheaper to import coal from South Africa, Columbia, or Russia than to support coalfield communities in Wales and Appalachia.

Taken at face value, international commerce seems to erode community self-determination, but can international connections also support local control? The answer may lie in the concept of internationalism, more specifically the idea of constructive internationalism as defined by David Unger in a paper for the World Policy Institute:

> Constructive internationalism sees us all living on one planet, with our primary international interest making that planet safer—from global warming, nuclear weapons, infectious diseases, and the widening inequalities that weaken democracy and help feed support for ideologies of hatred, xenophobia, and racism (Unger 2012).

We cannot escape the influence of the global economy, but we can work to actively shape our own local economies. At the same time, we can reach out to international communities that are facing similar challenges. Building power in coal communities involves facing a series of fears—fear of the unknown, the foreign, the future.

BY TOM HANSELL

My first experience underground was in Siberia. Sure, I had stuck my head in a few doghole mines in eastern Kentucky. However, entering these scraped-out mines involved crawling on hands and knees through black water into an opening less than a yard high. Needless to say, they did not hold a strong attraction for me, and I never ventured beyond the short reach of daylight.

In March 1999, I found myself riding in a bus across the frozen plains of Siberia to a mine entrance. Appalshop filmmaker Herb E. Smith and I were documenting a delegation of inspectors from the Mine Safety and Health Administration (or MSHA, part of the US Department of Labor) on a training mission to Russia and Ukraine. When we stopped, I looked for a familiar low portal and was surprised to see a six-foot-tall metal door in the side of the hill.

Our group entered the mine as easily as entering any other building, and I was momentarily blinded by the contrast between the snow-covered landscape outside and the dimly lit tunnel. As we turned on our headlamps, I could see the pathway sloping in front of us, alongside the conveyor belt that transported coal out of the mine. Before my eyes could completely adjust, Sasha, the mine foreman, started down the slope, urging us to follow.

At that time both Russia and Ukraine were emerging from Soviet rule and struggling with how to adapt coal mining and other industries from the production-based Soviet system to a market-based economy. The fatality rate in these countries was astonishingly high; only China killed more miners each year. Our delegation was a joint effort of the US State Department, the Department of Labor, and a private group called Partners in Economic Reform, or PIER. PIER was funded by the coal industry to encourage investment in the former Eastern Bloc. This exchange offered the Russians the expertise of American mine safety experts. In return, a representative of PIER was granted inside access to production statistics and other valuable information that allowed him to identify the most productive mines for his clients to invest in.

We had not traveled far down the slope of the tunnel when I slipped on the muddy floor. Out of instinct I grabbed the closest thing I could use to steady myself, which happened to be the beltline. Fortunately the mine manager had shut down the beltline during our passage into the mine. If it had been running as usual, my fingers would have been crushed between the coal-laden belt and its rollers. Straightening myself, and noticing the high-voltage power cables on the opposite wall, I quickly made a note to keep my hands by my sides, even if it meant falling.

This mine was in the Kuzbass, the vast Siberian coalfields near the city of Kemerovo. Looking at the map, I realized that we were closer to Mongolia than to Moscow. I was astonished to be walking upright alongside eight-foot-tall seams of coal that sparkled like stars in the beam of our headlamps. This mine's challenge was how to deliver their high-quality coal to markets that were many time zones away. When we visited mines in

Mine Safety Mission preparing to go underground in Ukraine. The author is third from the right.

Ukraine, the situation was very different.

We flew from Kemorovo to Donetsk, the heart of the Donbass coalfields that underlie the hotly contested border between Russia and Ukraine. From Donetsk, we traveled to Luhansk, where we were able to look across the Donets River into Russia. There we visited the Komsomolets (young communist) mine, named after the local group of youth who had resisted the Nazi occupation. The irony of Americans visiting a mine dedicated to young communists was not lost on our hosts.

The miners in the Donbass had been particularly hard hit by the breakup of the Soviet Union. Although the mines were close to international markets, coal had been extracted for more than a century. The remaining coal seams were thin and located deep underground, creating dangerous working conditions. The situation was so desperate that most miners were owed several months' back pay, and some miners worked only to produce the coal required to heat and power the massive concrete Soviet-era apartment complexes where they lived with their families. Suicides were common in these communities, and our hosts carefully placed barriers between the international visitors and the miners, afraid of what would happen if we were to come into contact.

The closest contact we made was through our translator. Valentin was a former English teacher who had turned to mining for the increase in pay. When the mine managers learned that he was one of the few Ukrainians who could understand technical mining terms in two languages, he was hired to help international visitors. Late one night in a bar, he shared his concerns about where the post-Soviet coal industry was heading:

"As we transition to the market economy, uneconomic mines are to be shut down, and they are the majority. Out of the three hundred coal mines in the Ukraine, about two hundred are to be closed. This creates social issues to be addressed—what to do with the laid-off miners?

"Right now there is about 750,000 people working in the industry, directly or indirectly. A mine as an enterprise took care of so many social assets: kindergarten, villages, hospitals, schools, and other stuff. So what do we do with them? When the mine closes, who will take over? The municipalities? They don't have funds either.

"Miners in this country are driven to the extreme point. They are not paid many months' back wages. And they don't care about these concepts of market economy or distribution economy. They are just not paid. That is the basic bottom line. They are at the brink of physical survival. They are not paid. How can they live? No money to buy food, to take care of their kids, that is very, very distressing.

"For the government to declare that we are striving to get to the market economy doesn't mean to implement that. Right now in Ukraine we have only slogans but no deeds. What is a market-oriented economy? Private business should be in place. Today we just have a few private shops, the basic industries are not privatized, and there is no plan or program. We are moving from a distribution economy to a market economy and we are stuck midway. What possible results can we have? Deterioration and regression."

People in the Donbass were very poor, certainly much poorer than in Appalachia, the poster child for American poverty. The newly independent Ukrainian government had no plan to transition coal and other industries from the Soviet system, which relied on production quotas, to a market-based system. As a result, the region was ripe for revolt, which arrived in 2014 when armed pro-Russian forces took over government buildings and declared the area a People's Republic.

In 2015, I was asked by the Organization for Security and Co-operation in Europe (an arm of the United Nations) to present my work on the After Coal project at a conference in Ukraine. The conference organizer confirmed that the United Nations peacekeepers in the region consistently identified the government's lack of support for laid-off miners and steelworkers as the chief cause of the conflict.

It is easy for Americans like me to point out the follies of Soviet-style communism. However, it is important to remember that the US State Department had a direct hand in promoting many of the free-market policies that ignored the needs of miners and steelworkers and sowed the seeds for the ongoing conflict in Ukraine. My government partnered with international energy companies to cherry-pick a few of the best mines for foreign investment. Part of the business plan was to bankrupt the competition, putting tens of thousands of miners out of work with no support for job training, relocation, or hope for the future. We may not have literally planted the seeds for revolt, but our actions fertilized those seeds and encouraged them to grow. It is extremely difficult to find ways to support international mining communities in a global economy, but the story of Ukraine illuminates the dangers of abandoning these places to the whims of the free market.

Meanwhile, back in Siberia, it was early in my travels and I was blissfully unaware of the global context surrounding this so-called mine safety mission. Mesmerized by the glittering coal seam, trying to follow the patter of the miners and mine inspectors and record the sounds of massive machines shearing wall-sized swaths of coal, I was surprised to find that our group had stopped at the bottom of the beltline that we had followed into the mine. This time the beltline was running.

The Mine Safety and Health Administration inspectors in our group paused and looked at each other, muttering concerns that I could not quite hear. Then one of the Russians climbed up onto a metal step and hopped onto the conveyor belt. Clearly, this was the way out. Through our interpreter, the MSHA staff quickly conferred with the Russians, explaining that riding belts was against safety regulations. After a brief exchange, the interpreter shrugged, stepped on the belt, and waved for us to follow. Herbie and I, holding our precious camera and sound equipment looked to the MSHA crew for direction. While we were waiting, Sasha, the mine foreman, stepped beside us, deftly lifted the camera from Herbie's hands and hopped on the belt, gleefully shouting "ooo-pa-pa!" as he disappeared into the darkness. Herbie and I looked at each other, shrugged, and followed, riding the beltline to daylight.

During the process of writing this book, on June 23, 2016, the United Kingdom voted to leave the European Union, beginning the process now known as Brexit. The Leave campaign that preceded the vote played on fear of immigrants, and the coalfields of south Wales supported Brexit by a narrow margin. Five months later, US voters elected Donald Trump president. Trump's Make America Great Again campaign received strong support from voters in former coal-mining and industrial regions, who were attracted by his promises to renegotiate international free trade deals in order to protect American workers and to crack down on illegal immigration. On both continents, fear of an uncertain future created support for isolationist policies. The unemployed blue-collar workers (including some laid-off miners) who supported the Trump campaign or the Brexit vote felt that the global economic system had left them behind—and they were right.

However, if we are to address legitimate concerns about workers who are left behind by a global economy, then shutting down borders and blaming foreigners is not a solution. We need to work across borders to develop systems of mutual support. In fact, closing borders can cut off important resources that have allowed communities to regenerate on their own terms. For example, several of the Welsh initiatives featured in the *After Coal* documentary, such as the DOVE Workshop and the Glyncorrwg Ponds and Mountain Bike Centre, have strategically used funding from the European Union to increase opportunities in the communities they serve.

In the Appalachian coalfields, fear of the federal government is often stronger than fears of foreign immigrants. Yet, federal agencies such as the Department of Agriculture, the Economic Development Administration, and the Appalachian Regional Commission often play an important role by distributing resources to high-priority areas such as coalfield communities. For example, most local governments in Appalachia do not have the resources to protect their water sources. Often, local officials turn to federal programs such as the Abandoned Mine Land Fund to run water lines to rural residents, reclaim mine sites, and provide a foundation for sustainable development. Clearly, the government should play an important role by providing access to clean water, which is a critical requirement for economic development.

In order to achieve a just and equitable transition from coal (and other fossil fuels), we need to develop global systems that do not leave working people behind, global systems that support local communities. These systems could be linked internationally, but in the words of Wendell Berry (2001), they cannot "export local products until

ROMANIA

INTERVIEW WITH GABRIEL AMZA BY TOM HANSELL

In October 2015 I traveled to the Carpathian Mountains of Romania to present the After Coal project at a conference titled "Appalachians/ Carpathians: Researching, Documenting, and Preserving Highland Traditions."

Like many Americans, I knew little about the Carpathian Mountains, the range that stretches across Central Europe, including parts of the Czech Republic, Slovakia, Poland, Hungary, Ukraine, and most of Romania. At the conference, biologist John Akeroyd from the ADEPT Foundation described the region as "the last of old-world Europe" and a hotbed of biodiversity. The fact that Romania was part of the Soviet-controlled Eastern Bloc until their 1989 revolution means that many rural areas are not developed. Horse-drawn carts and handmade haystacks are still a common sight in the countryside.

However, the Soviets also built immense industrial centers in parts of Romania. The Jiu Valley is a coal-mining region on the edge of the Carpathians. Under Soviet control, the largest mine employed tens of thousands of workers. After the 1989 revolution, the mines were privatized and workers lost their jobs. Declining coal markets and international competition have resulted in a plan to close all of the mines in the Jiu Valley over the next decade.

At the Appalachians/Carpathians conference I presented the *After Coal* documentary alongside Romanian photographer Gabriel Amza, who shared his ongoing documentary photography project Genus Loci—or spirit of place—which aims to define the quintessential spirit of the Jiu Valley. Genius Loci consists of beautifully haunting work that captures the despair and abandonment that many mining communities feel, without resorting to stereotypes. Gabriel Amza explained some of the complexities of his project to me:

TOM HANSELL: Tell me about the Jiu Valley.

GABRIEL AMZA: Historically, it used to be shepherding country, up until the industrial revolution, when coal was discovered in the area. Since then it's been coal country. First there were open-pit mines, and eventually they ended up being underground. People moved there from all over Europe. You had migrations during the Austro-Hungarian empire from as far as Russia, France, the Netherlands, Spain. You have all these diverse communities coming to the valley to get work in the mining industry.

In the 1920s the Petrila Mine, which is the largest and oldest in the area, was employing thirty thousand people. It was a whole town by itself. You know, twenty thousand people were put out of

Above: Tower at a mine site in Romania's Jiu Valley. Right: The Petrila mine in Romania's Jiu Valley.

work in Wales by Thatcher, but in the Jiu Valley you had twenty thousand people in one mine, and there were sixteen separate mines.

TOM HANSELL: That is an astonishing scale.

GABRIEL AMZA: Of course, in the communist era, the miner was considered one of the hardest workers and a respected pillar of society, but that only gets you so much when you're dying of black lung. One of the issues the area's confronting right now is that, in the 1990s, the Jiu Valley was responsible for somewhere around 95 percent of all work-related accidents in heavy industry in Romania because of the pressure to excel and to get as much coal out as cheap as possible. For people to keep their jobs, they would very regularly try and trick the safety tests. For example, the gas detectors would be set up next to the air vent, so it would detect even less of the deadly substances that occurred underground. Because of that, in my lifetime I have seen immense, massive, and very tragic mining accidents with hundreds of miners killed.

TOM HANSELL: What is life like now in the Jiu Valley?

GABRIEL AMZA: The mines are closing as we speak. The biggest one, the Petrila Mine, which had over fifty thousand workers in its heyday, will close later this month completely, losing the last five thousand jobs for the people of the community.

The situation right now is that the residents of the valley are poor. There are no other jobs in the community to replace those lost with the closing of the mines. You have a situation that is similar to what happened in Wales with the Thatcher reforms and the closing of the mines. It happened in Appalachia. It happened in Germany. It happened in the north of France. It happened in Holland as well. It's been happening in Sweden in the north. It's happened in Norway. It's been happening everywhere where you have these centers of singular industry, which then shifts out of economic need, and leaves people behind.

TOM HANSELL: So how do you address this situation through photography?

GABRIEL AMZA: I think coal is dirty. But, coal is an important part of our recent history. And the reason projects such as Genius Loci and After Coal are important is that just because coal is a dirty reality, it shouldn't be forgotten. We need to remember where these communities started. It's not always a pretty beginning, but it's a necessary part of the story.

Where do they go from here? It's the communities' own responsibility to create a new future. But it's up to us to document these things because other people aren't going to do it, and not everyone wants to remember the dirtiness of coal. Our role is to say: This has happened, there was suffering, there was strife, there was joy, there was happiness, there was beauty, and there was a way of life.

local needs have been met." Berry astutely observes that scarcity of resources is a major cause of anger and political unrest. Meeting local needs is an essential step to create peace and prosperity. Developing international connections to support coalfield communities on a path to self-sufficiency is a key concept to create a future beyond fossil fuels.

THE CHALLENGE OF SCALE

During the panel discussion following *After Coal*'s premiere at the Hay Festival in Wales, historian Dai Smith asked a difficult, yet essential, question about the scale of current efforts to regenerate coalfield communities. To paraphrase, he asked: "The material cultures that were created by coal have gone. Today, can the positive values that helped working-class people build power through solidarity actually

After Coal project advisor John Gaventa (left) speaking to Mair Francis at the 2016 Hay Festival.

be sustained by activities such as mountain bike riding, or tourism, or does there have to be something bigger or deeper?"

This question of how to keep the positive elements from the past in coalfield communities while living in a post-coal economy is essential for those of us who care about these places. Even if you do not particularly care about coalfield communities, it is still important that people in former mining regions participate in prosperity in order to keep a stable society. That said, there is plenty of evidence that supports the view that the world has changed too quickly for these places to catch up, that the culture developed in coalfield communities is a relic of the past.

However, an alternative view is that many lessons instilled in mining communities actually help us adapt to dramatic change. For example, coalfield residents have had to learn habits of resilience in order to adapt to the boom-and-bust cycles of coal markets. Even during boom times, mining families had to be ready to bounce back from mine explosions, floods, and other disasters that came along with this difficult and dangerous line of work. This collective resilience shines through at unexpected moments. At the Hay Festival, John Gaventa spoke about a deep desire to create a new future that he had witnessed in communities facing deindustrialization:

> If you go to the West Virginia coalfields, or Ohio, or the native, indigenous communities of Canada, they're all going through this global transition where those at the top have more and more, those in the poorer working-class communities have less and less. The global industrial economy is breaking down, and global inequality is growing. It's not just that transition from industrial to postindustrial, this transition pushes the haves and have-nots further apart.
>
> Out of those conditions, from the values of the past and the lingering culture, the next generations will have the

capacity to create something new. It will be very different, it won't be based on an industrialized working class. We don't know yet what it will be. But things are happening in these communities and in similar communities in many places around the world—in Latin America, and in Spain, poor parts of the United States. Some of these communities are beginning to link up, and they're beginning to talk about new kinds of solidarity economies.

A lot of these initiatives are small, but what do they give us? They give us an idea of how the world can be different. About how we can build our communities from within. I look at these projects that are emerging as actually teaching us about a new kind of economy of the future. And it's not just [in] Appalachia, it's not just [in] Wales, there are thousands of such examples all over the world.

In those spaces, we're seeing new kinds of economic structures being built. Structures that try to link culture and sustainability, economic development and use of land in new kinds of ways. So actually, I'm a bit optimistic that in this crisis, something new is being born and developed that will once again have an impact on the future and the bigger economy. So let's wait and see, see where it goes.

CREDITS

IMAGE CREDITS

All photographs, unless otherwise noted, are by Tom Hansell.

Page	Source
ii	Photo by Martin Shakeshaft (www.strike84.co.uk).
vi	Railway cargo cars carrying coal by satephoto / Shutterstock.
viii	Photo by Ben Rorick for After Coal.
11	Courtesy of the Center for Appalachian Studies at Appalachian State University.
18	Courtesy of the US Steel Collection in the Appalachian Archives at Southeastern Kentucky Community and Technical College.
19	Hand-drawn map by Pat Beaver. Photo by Geoff Charles, courtesy of the National Library of Wales.
27	Photo by Hywel Francis.
28	Photo by Suzanne Clouzeau for After Coal.
31	Courtesy of the W.L. Eury Appalachian Collection at Appalachian State University.
35	Courtesy of the Appalshop Archives.
38	Courtesy of the W.L Eury Appalachian Collection at Appalachian State University.
41	Hand-drawn map by Pat Beaver.
42	Courtesy of the US Steel Collection in the Appalachian Archives at Southeast Kentucky Community and Technical College.
45–46	Courtesy of the US Steel Collection in the Appalachian Archives at Southeast Kentucky Community and Technical College.
50	Courtesy of the US Steel and Benham Collections in the Appalachian Archives at Southeast Kentucky Community and Technical College.

53–54   Photos by Russell Lee, courtesy of the National Archives and
          Records Administration.

55   Aberfan disaster photo courtesy of the National Library of Wales.
          Buffalo Creek disaster photo courtesy of Appalshop.

56   Courtesy of the US Steel Collection in the Appalachian Archives at
          Southeast Kentucky Community and Technical College.

61   Wales photo by Pat Beaver for After Coal.
          Kentucky photo by Tom Hansell.

64   Photo by Martin Shakeshaft (www.strike84.co.uk).

66   Courtesy of the South Wales Miners' Library.

68   Courtesy of the Appalshop Archives.

72   Courtesy of After Coal.

77   Courtesy of After Coal.

88   Courtesy of After Coal.

90   Mair Francis photo by Tom Hansell.
          Strike photo by Martin Shakeshaft (www.strike84.co.uk).

93   Courtesy of After Coal.

101   Photo by Mair Francis for After Coal.

110   Courtesy of After Coal.

112–113   Courtesy of After Coal.

115   Courtesy of After Coal.

118   Courtesy of After Coal.

131   Photo by Ben Rorick for After Coal.

139–140   Courtesy of After Coal.

141   Photo by Ben Rorick for After Coal.

150   Courtesy of After Coal.

166   Photo by Suzanne Clouzeau for After Coal.

178   Photo by Angela Wiley for After Coal.

210–211   Photos by Gabriel Amza.

AFTER COAL FILM CREDITS

The *After Coal* documentary provided the idea for this book, and I believe that it is important to acknowledge the massive team effort that went into recording the interviews and visiting the locations featured in these pages. These credits provide a complete list of the people who made the documentary, and the entire After Coal project, possible.

Director: Tom Hansell

Producers: Patricia Beaver and Tom Hansell

Associate Producer: Angela Wiley

Film Crew: Suzanne Clouzeau, Herb E. Smith, Shawn Lind, Austin Rutherford, Kelly Davis, Rebecca Jones, Sally Rubin

Project Advisors: Richard Davies, Ronald Eller, Hywel Francis, Mair Francis, John Gaventa, Richard Greatrex, Frank Hennessy, Helen Matthews Lewis, Ronald Lewis, William Schumann, Victoria Winckler, Jack Wright

Production Assistants: Lauren Boone, Sarene Cullen, Samantha Eubanks, Austin Getzelman, Cheryl Laws, Skye McFarland, Baylor Rossi, Ben Rorick, Forrest Yerman

Funding provided by: Appalachian State University, Chorus Foundation, Kentucky Educational Television, West Virginia Humanities Council, United States Artists

Fiscal Sponsors: Southern Appalachian Labor School, Southern Documentary Fund

Special thanks to: Appalshop Media Arts Center, Center for Appalachian Studies at Appalachian State University, DOVE Workshop, Higher Ground of Harlan County, Kentuckians for the Commonwealth, Merthyr Tydfil College, University of Kentucky Appalachian Center, WMMT 88.7 FM, Woodrow Wilson Center

NOTES

1. These video recordings set the stage for John Gaventa's later work with the Highlander Research and Education Center in Tennessee, for which he received a MacArthur Fellowship, or "Genius Grant," and Oxfam in Great Britain, where he was made a member of the Order of the British Empire. After editing this documentary footage, Richard Greatrex became a successful cinematographer, who won a series of British Academy Film Awards and was nominated for an Academy Award for his work on *Shakespeare in Love* (1998). Helen Lewis authored important scholarship on labor and social justice and is considered the grandmother of the Appalachian Studies movement. Hywel Francis became a member of Parliament, representing the Aberavon constituency in south Wales from 2000 to 2015.

2. The story of the London gay and lesbian group that supported the Welsh miners during the 1984–85 strike is featured in the feature film *Pride* (2014), written by Stephen Beresford and directed by Matthew Warchus. The minibus that was eventually donated to the DOVE Workshop features prominently in the film.

3. The Big Sandy Power Plant in Rocky Adkins's district closed in 2015, replaced by natural gas (Waitkus 2015).

BIBLIOGRAPHY

Appalshop. 1994. Appalshop Mission Statement. Whitesburg, Kentucky.

Barret, Elizabeth. 1987. *Long Journey Home*: Appalshop Films. VHS.

Beckel, Michael. 2017. "Koch-Backed Nonprofit Spent Record Cash in 2012." The Center for Public Integrity 2013 [cited 21 May, 2017]. Available from www.publicintegrity.org/2013/11/14/13712/koch-backed-nonprofit-spent-record-cash-2012.

Berry, Wendell. 2001. "The Idea of a Local Economy." *Orion*.

Bevan Foundation. 2017. *About Us* 2015 [cited 23 January, 2017].

Brown, Jak. 2016. "Past Prime Ministers: Clement Attlee" 2012 [cited December 15, 2016]. Available from www.gov.uk/government/history/past-prime-ministers/clement-attlee.

Brown, John H. 2009. *The Valley of the Shadow: An Account of Britain's Worst Mining Disaster, the Senghennydd Explosion*. Port Talbot, West Glamorgan: Alun Books.

Caudill, Harry M. 1963. *Night Comes to the Cumberlands: A Biography of a Depressed Area*. Boston: Little, Brown and Co.

Coal Age News. 2012. "100 Years with Coal Age." *Coal Age*, September 14, 2012.

Commonwealth, Kentuckians for the. 2017. "About the ABF Conference" 2013 [cited June 30, 2017]. Available from www.kftc.org/abf/about.

Corbin, David Alan. 1981. *Life, Work, and Rebellion in the Coal Fields: The Southern West Virginia Miners 1880-1922*. Urbana, IL: University of Illinois Press.

Couto, Richard A. 1993. "The Memory of Miners and the Conscience of Capital: Coal Miners Strikes as Free Spaces." In *Fighting Back in Appalachia*, edited by Stephen L. Fisher. Philadelphia: Temple University Press.

Culver, J. and Mingguo Hong. 2016. "Coal's Decline: Driven by Policy or Technology?" *The Electricity Journal*, no. 29 (7).

Curtis, Ben. 2013. *The South Wales Miners 1964–1985*. Cardiff: University of Wales Press.

Davies, John. 2007. *A History of Wales*. London: Penguin UK.

Davies, John and Nigel Jenkins. 2008. *The Welsh Academy Encyclopedia of Wales*. Cardiff: University of Wales Press.

Desrochers, Daniel. 2016. "McConnell: It's 'Hard to Tell' if Ending 'War on Coal' Will Bring Back Jobs." *Lexington* (Kentucky) *Herald-Leader*. November 11, 2016.

Donohue, Jean. 1990. *From the Shadows of Power*. Cincinatti, OH: Media Working Group.

Eller, Ronald D. 1982. *Miners, Millhands, and Mountaineers: Industrialization of the Appalachian South, 1880–1930*. Knoxville, TN: University of Tennessee Press.

Eller, Ronald D. 2008. *Uneven Ground: Appalachia since 1945*. Lexington, KY: University Press of Kentucky.

Energy and Environment Cabinet, Commonwealth of Kentucky. 2016. "Kentucky Coal Facts 2016." Frankfort, KY.

Francis, Hywel. 2009. *History on Our Side: Wales and the 1984–1985 Miners' Strike*. United Kingdom, Cardigan, Wales: Parthian Books.

Francis, Hywel. 2010. Keynote Speech. In *Appalachia and Wales: Coal and After Coal*. Appalachian State University, Boone, NC.

Francis, Hywel and David Smith. 1998. *The Fed*. Cardiff: University of Wales Press.

Gildart, Keith. 2011. "The Women and Men of 1926: A Gender and Social History of the General Strike and Miners' Lockout in South Wales." *Journal of British Studies*, no. 50 (3): 758–59.

Hansell, Tom. 2016. *After Coal*. Boone, NC: Appalachian Journal. DVD.

Hardt, Jerry. 2013. "Appalachia's Bright Future." In *April Conference to Focus on Creating Economic Opportunities*: Kentuckians for the Commonwealth.

Hitt, Mary Ann. 2015. *Move Beyond Coal—United States Narrative*. Washington, D.C.: Sierra Club.

IEA. 2013. "Medium Term Coal Market Report." Paris, France: International Energy Agency.

Lewis, Anne. 1986. *Mine War on Blackberry Creek*. Whitesburg, KY: Appalshop Films.

Lewis, Helen M. 1976a. *Dai Francis interview*. Boone, NC: W.L. Eury Appalachian Collection, Appalachian State University.

Lewis, Helen M. 1984. "Industrialization, Class, and Regional Consciousness in Two Highland Societies: Wales and Appalachia." In *Cultural Adaptation to Mountain Environments*, edited by Patricia D. Beaver and Burton L. Purrington, 50–70. Athens, VA: University of Georgia Press.

Lewis, Helen Matthews. 1976b. *The Welsh Tapes*. Boone, NC: W.L. Eury Appalachian Collection: Appalachian State University.

Lewis, Ronald L. 2008. *Welsh Americans: A History of Assimilation in the Coalfields*. Chapel Hill, NC: University of North Carolina Press.

Maher, Kris. 2014. "Mine Workers Union Shrinks but Boss Fights On." *Wall Street Journal*, January 9, 2014.

Massey, Doreen. 1994. *Space, Place and Gender*. Minneapolis, MN: University of Minnesota Press.

McAteer, Davitt. 2007. *Monongah: The Tragic Story of the 1907 Monongah Mine Disaster, the Worst Industrial Accident in US History*. Morgantown, WV: West Virginia University Press.

McLean, Ian and Martin Johnes. 2009. *Aberfan: Government and Disasters*. Cardiff: Welsh Academic Press.

Mineworkers, National Union of. 1965. *The Miner*.

National Mining Association. 2007. "Most Requested Statistics—U.S. Coal Industry."

Parliament, United Kingdom. 2016. [cited December 16, 2016]. Available from www.parliament.uk/about/living-heritage/transformingsociety/livinglearning/19thcentury/overview/coalmines.

Pavlovich, Joe. 1999. MSHA safety training. Donetsk, Ukraine.

Pickering, Mimi. 1975. *Buffalo Creek: An Act of Man*. Appalshop Films: Whitesburg, KY.

Politico. 2017. "Kentucky Presidential Results 2012." Politico 2012 [cited June 30, 2017]. Available from https://www.politico.com/story/2012/11/election-results-2012-by-state-083320.

Portelli, Alessandro. 2012. *They Say in Harlan County: An Oral History*. New York: Oxford University Press.

SALS. 2017. *About the Southern Appalachian Labor School*. Southern Appalachian Labor School 2016 [cited February 26, 2017]. Available from http://sals.info.

Scargill, Arthur. 1983. *Miners in the Eighties*, edited by Yorkshire Area National Union of Mineworkers. Barnsley, England: Albert Taylor and Sons.

Social Enterprise Alliance. 2017. *What Is Social Enterprise?* 2017 [cited June 3, 2017].

The West Virginia Encyclopedia, e-WV. 2017. *"Joy Loading Machine."* e-WV: The West Virginia Encyclopedia., March 5, 2012 [cited October 23, 2017]. Available from www.wvencyclopedia.org/articles/1060.

Tower Regeneration. 2017. "History of Tower Colliery." Available from http://www.towerregeneration.co.uk/history/.

Unger, David C. 2012. "A Better Internationalism." *World Policy Journal*, no. 29 (1). doi: 10.1177/0740277512443809.

USEIA. 2015. "What Is U.S. Electricity Generation by Energy Source?" www.eia.gov/tools/faqs/faq.cfm?id=427&t=3.

Vaughan, Adam. 2016. "Britain's Last Coal Power Plants to Close by 2025." *The Guardian*.

Waitkus, Dave. 2017. "Kentucky Power Remembers Big Sandy Power Plant as Transition to Natural Gas Begins." American Electric Power 2015 [cited May 23, 2017]. Available from https://aepretirees.com/2015/06/19/kentucky-power-remembers-big-sandy-power-plant-as-transition-to-natural-gas-begins.

Weeks, Phillip. 2004. *Coal Man's Secret Diaries*. BBC Radio Wales. Radio Report.

Women's Bureau (Department of Labor). 1985. *The Coal Employment Project— How Women Can Make Breakthroughs into Nontraditional Industries*. Washington, D.C.: United States Department of Labor.

INDEX

A